W9-BJZ-987

Physician-Assisted Death

Biomedical Ethics Reviews

Edited by

James M. Humber and Robert F. Almeder

Biomedical Ethics Reviews • 1993

Physician-Assisted Death

Edited by

James M. Humber,
Robert F. Almeder,
and Gregg A. Kasting
Georgia State University, Atlanta, Georgia

Humana Press • Totowa, New Jersey

The Library of Congress has cataloged this serial title as follows:

Biomedical ethics reviews—1983– Totowa, NJ: Humana Press, c1982–

v.; 25 cm—(Contemporary issues in biomedicine, ethics, and society)
Annual.
Editors: James M. Humber, Robert F. Almeder, and Gregg A. Kasting.
ISSN 0742-1796 = Biomedical ethics reviews.

1. Medical ethics—Periodicals. I. Humber, James M. II. Almeder, Robert F. III. Series.

[DNLM: Ethics, Medical—periodicals. W1 B615 (P)]

R724.B493   174'.2'05—dc19   84-640015
AACR2 MARC-S

Contents

Preface

Physician-Assisted Death is the eleventh volume of *Biomedical Ethics Reviews*. We, the editors, are pleased with the response to the series over the years and, as a result, are happy to continue into a second decade with the same general purpose and zeal.

As in the past, contributors to projected volumes have been asked to summarize the nature of the literature, the prevailing attitudes and arguments, and then to advance the discussion in some way by staking out and arguing forcefully for some basic position on the topic targeted for discussion.

For the present volume on *Physician-Assisted Death*, we felt it wise to enlist the services of a guest editor, Dr. Gregg A. Kasting, a practicing physician with extensive clinical knowledge of the various problems and issues encountered in discussing physician-assisted death. Dr. Kasting is also our student and just completing a graduate degree in philosophy with a specialty in biomedical ethics here at Georgia State University. Apart from a keen interest in the topic, Dr. Kasting has published good work in the area and has, in our opinion, done an excellent job in taking on the lion's share of editing this well-balanced and probing set of essays. We hope you will agree that this volume significantly advances the level of discussion on physician-assisted euthanasia.

Incidentally, we wish to note that the essays in this volume were all finished and committed to press by January 1993. In February 1993, The Netherlands passed a law permitting physician-assisted death, and is now actively debating legislation permitting involuntary euthanasia under well-defined conditions.

James M. Humber
Robert F. Almeder
Gregg A. Kasting

Contributors

John C. Fletcher • Center for Biomedical Ethics, University of Virginia, Charlottesville, Virginia

Gregg A. Kasting • Department of Philosophy, Georgia State University, Atlanta, Georgia

Diane E. Meier • Mount Sinai Medical Center, New York, New York

Franklin G. Miller • Center for Biomedical Ethics, University of Virginia, Charlottesville, Virginia

G. Steven Neeley • 9452 Meadow Ridge Drive, Cincinnati, Ohio

Evelyne Shuster • Department of Veterans Affairs Medical Center, Philadelphia, Pennsylvania

David C. Thomasma • Medical Humanities Program, Loyola University of Chicago Medical Center, Maywood, Illinois

Introduction

Diane E. Meier's chapter, "Doctor's Attitudes and Experiences with Physician-Assisted Death: A Review of the Literature," introduces the topic of physician-assisted dying. Meier begins by noting that codes of professional conduct prohibit mercy killing by physicians and that, currently, doctors cannot legally practice active euthanasia in the United States. Next, Meier discusses some of the reasons why there are challenges to these established constraints on euthanasia and physician-assisted death. This discussion is followed by a brief section in which key terms (e.g., "voluntary euthanasia," "assisted suicide") are defined and distinguished. Meier then reviews the arguments that have traditionally been offered for and against voluntary euthanasia and physician-assisted death, and contends that in the absence of further empirical data none can be taken to be truly decisive. In the final two sections of her paper, Meier examines public opinion polls and surveys of physicians' opinions and practices concerning euthanasia and physician-assisted death. Although Meier holds that the general public exhibits substantial support for easing restrictions on physician-assisted death and euthanasia, she also points out that all surveys of physicians' opinions and practices are flawed. In the end she concludes that we must have a more definitive study of physicians' attitudes and actual practices before making any substantive changes in public policy.

In "The Nonnecessity of Euthanasia," Gregg Kasting examines one of the arguments that opponents of legalized assisted death frequently present, one that has appeared in the biomedical ethics literature for over a decade. The "nonnecessity argument" is, namely, that euthanasia and physician-assisted suicide are not necessary because physicians are able to control pain adequately with presently available medicines and techniques. Kasting argues first, that this empirical assertion is not borne out by the medical data; second, that a weaker form of the argument is extensionally unsound based on the scope of terminal illnesses and the suffering they produce; and third, that the argument is logically unsound in that it ignores the many other forms of suffering besides pain. Kasting concludes that suffering is an unfortunate, but present, reality for many dying patients and that the belief that assisted death is simply unnecessary is not a true depiction of medical care available from even the best of clinicians at the present time.

In "The Constitutionality of Elective and Physician-Assisted Death," G. Steven Neeley argues that suicide should be viewed as a constitutional right. Neeley begins by examining the basic tests the Supreme Court has used to determine when it is proper for the government to interfere with individual liberty. He then examines the right to privacy as it has been recognized by the Supreme Court and in common law. Although Neeley acknowledges that some courts have appealed to the right to privacy to allow patients to refuse or suspend artificial life support, he notes that recent Supreme Court decisions have limited the right to privacy, and expresses doubt that this right will be extended so far as legally sanctioned suicide. Next, Neeley examines the notion of "fundamental human right," and argues that the right to suicide should be viewed as such a right. Finally, Neeley contends that recognition of suicide as a fundamental human right would not lead to abuse, for states could still place restrictions on the exercise of that right.

In "Physician-Assisted Suicide and Active Euthanasia," Franklin Miller and John Fletcher defend the view that both physician-assisted suicide (PAS) and physician-performed euthanasia (PPE) should be legalized in the United States. The authors begin by taking note of various cultural signs indicating that US citizens would not be shocked by a public policy that endorsed experimentation with PAS and PPE. Next, Miller and Fletcher argue that:

1. Appeals to patient self-determination and relief of human suffering can justify legalization of PPE for some terminally ill patients;
2. Those who believe that PAS is morally superior to PPE and that only PAS should be legalized are wrong, for PAS and PPE are roughly morally equivalent; and finally
3. Both PAS and PPE should be legalized in the United States, albeit with certain safeguards.

Miller and Fletcher insist on "safeguards" in (3) because they recognize that there are dangers inherent in legalizing PAS and PPE. To prevent or minimize these dangers, the authors argue that PAS and PPE should be permitted only for those patients who are competent, terminally ill (i.e., with a prognosis of no more than six months to live), and who voluntarily request to be killed. To ensure that competence, terminal illness, and voluntariness are truly present in all instances in which patients are euthanized or assisted in suicide, Miller and Fletcher insist that each case of PAS and PPE be subject to prior review by a specially constituted committee.

David Thomasma presents the core of the current discussion on medically assisted death in "The Ethics of Physician-Assisted Suicide." He discusses the central questions that we grapple with at this point: Is it morally permissible for individuals to end their lives when they no longer wish to go on living or suffering? What is the nature of the patient–doctor relationship, and does it include a responsibility or option to assist in suicide and euthanasia? What will be the impact on society if we legalize these acts? Does not beneficent intention make assisted death just as morally permissible as withholding or withdrawing treatment? Thomasma points out some of the pitfalls we might encounter if we continue to override patient autonomy and argues that, though we must never overlook the history of Nazi Germany, a descent down the slippery slope to involuntary euthanasia is not inevitable. In the end, Thomasma presents what he believes are the six strongest arguments for, and the five best arguments against, legalizing physician-assisted suicide. Though he feels there are sound reasons for permitting assisted suicide, the author urges that we first increase expenditures on improving care for the dying.

In "Life, Death, and the Pursuit of Destiny: The Case of Thomas Donaldson," Evelyne Shuster discusses the issue of physician-assisted death as it relates to the case of Thomas Donaldson. Donaldson was a terminally ill patient who suffered from an inoperable brain tumor. He asked the court for a declaratory judgment giving him permission to be killed and frozen in cryogenic suspension in the hope that he could be brought back to life when a cure was found for his condition. The court interpreted Donaldson's request as nothing more than a petition for the right to be given active euthanasia, and so ruled against Donaldson. Shuster argues convincingly that the Donaldson case presented the court with an appealing opportunity to further the public debate over the propriety of physician-assisted death, and that the court failed to take advantage of that opportunity. As Shuster sees it, Donaldson was not simply asking to be euthanized; rather he was asking to be "killed" in order to live in the future. For Shuster, then, Donaldson was acting in accordance with the state's interest in preserving life, and the court's decision not to take this into account did little to add to our knowledge concerning the propriety (or impropriety) of physician-assisted death. Furthermore, Shuster contends that an examination of the Donaldson case and its disposition by the court make at least two things evident: (1) We need much more public debate in order to determine whether the need for physician-assisted death is real, and (2) as a culture, we find quick action—even futile action—preferable to long, sustained efforts to overcome intractable problems; thus, if our society eventually approves of physician-assisted death, this may simply be because we find it a "quick fix" and easier to accommodate than continually struggling with alternatives such a long-term home care, hospices, and comprehensive comfort care.

Doctors' Attitudes and Experiences with Physician-Assisted Death

A Review of the Literature

Diane E. Meier

Introduction

From the time of Hippocrates, codes of professional ethics have prohibited mercy killing by doctors, and it remains illegal for a physician to actively assist in the death of a patient. In recent years, however, many social forces have combined to lead to heightened public and professional debate about the morality and legality of physician-assisted dying. In Washington State and California, for example, proponents of physician-assisted dying gathered sufficient signatures to place initiatives on recent ballots calling for physician aid in dying under carefully regulated and defined circumstances.[1] Both initiatives were defeated by narrow margins after extensive media campaigns financed in part by right-to-life organizations.

These efforts to challenge constraints on euthanasia have grown out of multiple factors changing the nature of the practice of medicine and altering the relationship between doctor and patient. Before antibiotics and high-technology medicine, doctor provided only human comfort and relief of suffering to the dying. Today the medical armamentarium is often successful at prolonging life in the face of many previously rapidly fatal illnesses. It is apparent that patients have not always experienced this shift in the capacities of modern medicine as an unalloyed good.[2-4] The increasing expression of anxiety that some have described[2-6] when contemplating debilitating and terminal illness evolves, in part, from the fear that the life prolonged by medical technology will be one of pain and dependency, and that the patient will have little or no control over the application of this technology.

Reasons for this loss of faith in the benefits of modern medical technology are not well understood but may be influenced by the fact that in recent years more people have been exposed to the effects, both good and bad, of modern medicine. Those who have witnessed an uncomfortable or painful delay of a family member's death by medical interventions may want the option of an alternative mode of dying for themselves. Massive media attention to dramatic right-to-die cases, such as Cruzan[7] and Quinlan,[8] has influenced others to question the merits of using medical technology to achieve the maximum duration of survival when quality of life is severely compromised.[1-12] The influence of this technology imperative on the practice of medicine, whereby the existence of an intervention with any possibility of benefit seems to mandate its use, has led to increased efforts to assess the outcomes of new technologies in order to avoid inappropriate application to patients with little likelihood of benefit.[13]

The public's desire for more control over the circumstances of death has also manifested in legislative[1] and case-law[7-9,11] efforts to legalize and disseminate the advance-directive and proxy decision-maker processes, whereby patients are enabled to exert control over health care decisions even after their decisional

201-436-0492 → cell

Jean

capacity is lost.[14,15] More medical education in the science of pain management and terminal care, and in identifying when these interventions become appropriate, has also been encouraged[6] as a means of lessening the suffering of dying persons. It is not known whether these efforts will lead to a greater willingness by clinicians to withhold or withdraw life sustaining therapies under those circumstances where the burdens of continued treatment outweigh the benefits of a longer life for an individual patient. However, even if terminal care and pain control techniques are improved and greater respect is accorded to patients' rights to control their own treatments, there are still some individuals for whom the burden of suffering during terminal illness outweighs the benefit of any additional length of life. Some physicians and ethicists have argued that for these persons the physician's compassionate obligation to relieve suffering should also include the option of assisted suicide and voluntary euthanasia.[16–23] The recent highly publicized case of assisted suicide for "Diane," a patient of Timothy Quill with acute leukemia,[18,23] suggests that the public is beginning to share this view: A grand jury declined to indict Dr. Quill for knowingly prescribing the barbiturates "Diane" used to kill herself, an act that is clearly illegal under New York State law. Several public opinion polls[24–29] have also indicated substantial public support for the option of physician-assisted dying.

Definition of Terms

Confusion over terminology often obscures the central issues in debates over euthanasia and assisted suicide. Euthanasia is defined as an easy death or means of inducing one and as the act or practice of putting to death persons suffering from incurable conditions or diseases.[30] If the act is undertaken at the explicit request of a competent patient, it is defined as voluntary euthanasia. In contrast, involuntary euthanasia—the act of killing someone without his or her explicit request—is clearly immoral and is fundamentally different from the problem raised by patients with intact mental capacity who request assistance in dying. Implicit in the definition of voluntary euthanasia is the recognition that death is in the best interest of the patient requesting assistance, as assessed by the patient, that his or her pain and suffering outweigh the benefit of any additional duration of life, and that the patient is physically incapable of suicide (e.g., cannot swallow pills). Assisted suicide, in contrast, is defined as the act of making the means of suicide (such as a prescription for barbiturates) available to a patient who is otherwise physically capable of suicide. The crucial difference between voluntary euthanasia and assisted suicide is that in voluntary euthanasia, the final act is performed by someone other than the individual who wishes to die.[17] The terms physician-assisted dying or physician-aid-in-dying generally refer to both voluntary euthanasia and physician-assisted suicide.

Arguments Pro and Con

The case against physician-assisted death is based on its implications for public policy as well as its impact on the moral integrity of the medical profession.[31-32] The public policy issues express the slippery slope concerns that abuses are more likely to occur in a permissive milieu where euthanasia is legal than in one where it is not. Several investigators have alleged that involuntary euthanasia occurs with alarming frequency in The Netherlands, where an informal policy permits voluntary euthanasia and assisted suicide.[32-36] Other mechanisms by which indications for physician-assisted suicide could be inappropriately expanded include permitting proxies to decide for euthanasia on behalf of incompetents (an indication permitted by the failed California Death with Dignity ballot initiative of 1988); subtle pressuring of chronically ill or dying patients to request euthanasia because of economic or emotional hardships posed to others by their illness;

and similar subtle societal or institutional coercion of vulnerable or disenfranchised groups, especially persons without health insurance.[32] In an era of aggressive cost containment efforts, these types of economic pressures on vulnerable patients should not be underestimated. Procedural safeguards could not guarantee that such abuses could be prevented. An easing of constraints on physician participation in euthanasia might also distract needed physician attention from efforts to improve pain control, terminal care, and respect for patient wishes for care at the end of life. Finally, it might be more difficult for patients to entrust their lives to doctors known to participate in euthanasia[31,32,37] because such actions, regardless of how noble the intention, could destroy the identity of medical professionals as healers with strong regard for the value and sanctity of life.

Arguments in favor of voluntary euthanasia and physician-assisted suicide usually turn on the patient's right to autonomy[22] and the physician's compassionate obligation to relieve suffering.[17,18,20] Proponents argue that even with the most sophisticated management of pain and the discomforts of terminal illness, severe intractable physical and existential suffering persists for many patients and cannot always be eliminated or reduced to tolerable levels utilizing modern techniques.[18] Other elements of human suffering, such as losses of functional capacity and independence, represent indignities that are unacceptable to some persons. Knowing that euthanasia or assisted suicide is an option can return a measure of control and independence to a dying person with these concerns.[18,19] Further, physicians are bound to care when cure is no longer possible, and that caring function rests on the compassionate relief of suffering[6,20,38] for terminally ill patients. Finally, when the primacy of patient autonomy legally permits a decision to forego life-sustaining treatment in the United States, a decision which will surely lead to death, the justification for the strong legal distinction between direct killing and allowing to die (active vs passive euthanasia) has often been challenged.[21,22]

The Need for Data

What has not been addressed in the foregoing debate is the true frequency of requests for and acts of physician-assisted dying in US clinical practice. It is possible that there are currently very few patients approaching doctors for help with dying. The highly publicized cases of recent months[1,6,18,25,37] may represent only isolated deviations from the norm. If this is the case, then current strict legal proscription of euthanasia may be effectively preventing abuses and encouraging appropriate terminal care and optimal pain control practices. Conversely, these highly publicized cases may represent the proverbial tip of the iceberg: Perhaps physicians are frequently confronted with such requests from genuinely suffering and mentally competent patients. It is also possible that physicians may accede to such requests more regularly than is now appreciated and that some fraction of these unpublicized cases may involve the types of abuses that are feared by opponents of euthanasia (for example, the "It's Over, Debbie"[38] case). If this is found to be the case then, at the very least, more physician education on ethical approaches to end of life decisions would be clearly indicated. Furthermore, proponents of euthanasia could also argue that limited legalization of physician-assisted dying with imposition of strict procedural safeguards might actually reduce the frequency of inappropriate or involuntary euthanasia. Thus, development of educational objectives, clinical practice guidelines, and public policy depends to a large degree on the availability of factual information on the true frequency and characteristics of requests for and acts of physician-assisted dying, data that are not currently available.

Why are there so little reliable data on the frequency of patient requests for and physician participation in acts of voluntary euthanasia or assisted suicide? Few such questionnaire or survey efforts have been made, possibly because of the sensitive religious, moral, and legal implications of the problem and possibly because of concerns about uncovering and exposing practices that some feel are best left between doctor and patient. The sensitive nature of the issue makes it difficult to rely on the honesty of responses from physicians concerned about their legal vulnerability if identified. Thus, it is likely that the prevalence of physician-assisted dying in such surveys would represent an underestimate since many doctors would be unlikely to admit to their actions. Those surveys that have been completed and published are plagued by problems of poor generalizability, small sample sizes, confusing or misleading wording, and low response rates with attendant respondent bias (persons favorably disposed to, or with strong feelings for or against, the survey issue may be more or less likely to respond). In spite of these methodologic weaknesses, available surveys do suggest a substantial interest in the problem of physician-assisted dying, on the part of both the general public and selected groups of doctors.

Public Opinion Polls

Surveys of the general public have shown substantial support for easing of sanctions on voluntary euthanasia and physician-assisted suicide. In 1990 the National Opinion Research Center reported[24] that 60% of a representative sample of US citizens felt that a person with incurable disease has the right to end his or her life (up from 49% the year before), and 72% felt that doctors should be able to end the lives of the hopelessly ill at the request of the patient. A 1990 *New York Times*/CBS News Poll[25] found that 53% of 573 nationwide respondents in a telephone interview felt that doctors should be allowed to assist an ill person in taking his or her own life. A 1990 Roper Poll commissioned by The Hemlock Society[26] involving face-to-face interviews with a representative sample of 2000 US adults, found that 63% felt that physician-assisted dying should be legalized. A 1991 Roper Poll/ Hemlock Society survey of a representative sample of 1500 adults (in California, Oregon, and Washington State) found that 68%

believed that doctors should be legally allowed to assist in the death of a person who has a painful and distressing terminal disease.[27] A 1991 *Boston Globe*/Harvard poll of a national sample of 1311 adults found that 64% favored physician-assisted suicide and euthanasia for terminally ill adults who request it.[28] A 1988 gallup survey of members of the American Bar Association found that 56.8% of lawyers polled supported legalization of active euthanasia for terminally ill patients who had stated the wish to die, with 51.3% stating that they would ask their own doctor for a lethal injection if they were hopelessly ill and in great pain.[29] These data suggest that a consistent majority of adult American respondents support some form of physician-assisted suicide for terminally ill patients.

Physician Opinions and Practices

Several surveys of physicians' attitudes and practices with regard to physician-assisted dying have been reported in recent years. Their results are herein described and critiqued in chronological order.

A 1987 Australian university-based anonymous survey[39] of 2000 randomly selected doctors yielded among the highest response rates (46%) of any recent physician survey on this topic. Forty-eight percent of the respondents had requests from patients to hasten their death and only these respondents were queried further about their actual clinical practices. Of these 354 doctors, 75% discussed the issue with patients' families or close friends, and 93% thought these requests were sometimes rational. Twenty-nine percent of the 354 admitted to taking active steps to bring about the death of a patient who had asked them to do so. Of the 71% who had not taken such steps, a majority (65%) rejected the request at least in part because of its illegality. Among those who had participated in voluntary euthanasia, a majority had done so more than once (80%) and most (96%) felt that, in retrospect, they had done the right thing. There were no clear differences in the practice of voluntary euthanasia based on gender or specialty, although fewer Catholic and Protestant respondents reported hastening death than other groups. The overall group of 869 respondents was asked about their attitudes toward voluntary euthanasia: 62% approved of the practice; 59% thought it would be a good thing to duplicate the practice of The Netherlands in this regard; 60% wanted legalization of voluntary euthanasia; and 40% said they would practice active voluntary euthanasia if it were legal. These data suggest a high level of physician support for legalization of voluntary euthanasia in Australia, and a substantial subset who freely admitted to the practice in the past (48% of whom signed their questionnaires, thus declining anonymity). Weaknesses of this study include the small absolute numbers of respondents, the possibility that Australian medical practice is fundamentally different from that of the United States and thus not applicable, and the low response rate.

A 1987 British telephone survey[40] commissioned by the British Voluntary Euthanasia Society and conducted by the National Opinion Poll Market Research Ltd. found that 30% of the British general practitioners contacted agreed with legalization of voluntary euthanasia, and 35% would consider such a practice for their patients in the future. Weaknesses of this study include its conduct over the telephone, precluding anonymity, the survey's association with a euthanasia advocacy organization, which may alter or diminish physician responses in other ways, the small numbers polled (150 respondents), and the possible lack of applicability to US physicians. The sampling frame and response rate were not described.

In the United States, several attempts to quantitate the practice of and attitudes toward voluntary euthanasia have been published or released. In 1987, the National Hemlock Society (a euthanasia advocacy organization) conducted an anonymous mail survey[41] of 5000 California physicians who were members of the American Medical Association and primarily in specialties of general practice and internal medicine. The most serious deficiency with this survey was that only 12% of the physicians receiving the questionnaire returned it. No followup attempts to increase response rate were reported, but reluctance to respond to a survey from an organization that strongly advocates voluntary euthanasia is probably a major factor responsible for the poor response rate. Among respondents, 57% had been asked by a patient to hasten death, usually because of terminal illness with persistent pain. Nearly 23% of respondents admitted to taking steps to hasten death in response to a patient request, the majority (81%) more than once. Of doctors who rejected such patient requests, 80% cited illegality as at least part of their reason for refusing. Over two-thirds of respondents felt that voluntary euthanasia under carefully defined circumstances should be legalized, and 51% of these said they would practice physician-assisted dying if it were legal. It is likely that a higher proportion of physicians sympathetic to the aims of the Hemlock Society returned the survey than those who were either neutral about or disagreed with the Society's position. Thus, these figures may overstate the frequency of voluntary euthanasia, and the attitudinal data are probably similarly skewed. Other major problems with this survey include its conduct in a state subject to extensive publicity for a 1988 ballot effort to legalize physician-assisted dying (and literature for this ballot was actually attached to the physician survey). The questionnaire language contains many ambiguous words and phrases (for example, the phrase "active steps to end someone's life" used in this survey is subject to highly variable interpretation), and the sample is not representative. Thus, this survey provides little information about the true prevalence of and attitudes toward physician-assisted dying.

In 1988 The Center for Health Ethics and Policy at the University of Colorado mail-surveyed[42] all licensed physicians in the state of Colorado ($n = 7095$) regarding many aspects of life sustaining treatment, including voluntary euthanasia and physician-assisted suicide. Thirty-one percent of those surveyed responded.

Among respondents, 37% admitted to giving pain medication with the effect (whether intended or not) of shortening the patient's life, 60% have cared for patients for whom they felt active euthanasia would be justified, and, of these, 59% would be willing to administer a lethal dose if it were legal to do so. Four percent admitted to assisting patients in stockpiling a lethal dose of medication for purposes of suicide. This study suggests that about one-third of Colorado's physician respondents to this survey are in favor of physician-assisted dying under defined circumstances, and that a substantial minority of these respondents had hastened a patient's death. As with the surveys described above, the major weakness in this survey is the low response rate, because it is likely that there is a tendency for those favorably disposed toward physician-assisted dying to respond at either higher or lower frequency than others. Another weakness involves the use of ambiguous and undefined terms, such as active euthanasia, which may be understood in variable ways by respondents.

In 1988 the San Francisco Medical Society anonymously surveyed[43] the opinions of its membership about euthanasia. Of 1743 surveys mailed, 39% or 676 were returned. The demographics of the respondents were representative of the overall membership. Seventy percent of respondents felt that terminally ill patients should have the option of active euthanasia (not defined), 54% felt physicians should administer the lethal dose, and 45% would accede to a patient's request for euthanasia if it were legal. Responses did not differ by gender, specialty, or age of physician, but Catholics were significantly less likely than other religious groups to support a euthanasia option. Weaknesses of this study include unclear wording and lack of definition of terms (such as the failure to provide a clear definition of the term active euthanasia), low response rate and attendant respondent bias, and small numbers. Finally, the survey was mailed at a time of extensive publicity for the California Death with Dignity ballot initiative, which would also be likely to bias the physician's understanding of and responses to the survey.

A 1991 anonymous survey conducted by *Physician's Management* magazine[44] was mailed to 2000 primary care physician-subscribers, of whom 489 (24%) responded. Among respondents, 3.7% admitted to deliberately providing information to patients or family members regarding means of suicide; 9.4% said they had "deliberately taken clinical action that directly caused a patient's death"; and 45% said they had taken clinical action that "indirectly" caused a patient's death. Thirty-four percent favored legalization of euthanasia. The major weaknesses in this data are the poor response rate (and associated respondent bias) and the use of highly ambiguous phrasing for the questions on physician aid in dying. For example, many respondents may have viewed removal of life-sustaining therapies such as ventilators to be a "direct" act causing a patient's death, and may not have understood the use of the term "direct" in the survey to refer to active euthanasia (i.e., an act solely intended to shorten a person's life). The source of the respondents (subscribers to the magazine) is not representative, and the sampling method was not described.

A 1991 survey of all 600 members of the Academy of Hospice Physicians yielded a 35% response rate and found that whereas respondents were overwhelmingly in favor of a patient's right to forego life sustaining treatment, only 5% agreed with providing a lethal prescription dose of medication (assisted suicide) and only 1.4% supported active euthanasia for a competent terminal patient with uncontrolled pain.[45] These figures rose to 23% if assisted suicide, or 14% if active euthanasia, were decriminalized in association with appropriate limitations and safeguards on the practice (the latter questions were directed to a subset of 70 of the original respondents, queried at a later date). This subset of 70 respondents ranked the importance of arguments against legalizing active euthanasia as follows: First in importance was diversion of attention away from optimizing palliative/comfort care skills and measures; second, subversion of the healing role of the doctor; third, moral unacceptability of ever actively causing the death of another person; and ranked fourth and last in importance, risk of slippery slope degeneration into involuntary euthanasia and other abuses. Problems with this data include the probable vested interest of respondents in opposing easing of sanctions against physician-assisted suicide because it would potentially diminish numbers of hospice patients, the poor response rate, and the small number of physicians queried on the impact that legalization might have on their attitudes.

A 1991 Dutch study of euthanasia practices in The Netherlands utilized anonymous questionnaires mailed to the general practitioners of 6942 deceased patients deemed to have died after some type of end-of-life medical decision-making.[36] These questionnaires were returned at a 76% response rate. In addition, a stratified random sample of 405 physicians caring for the majority of patients who died in The Netherlands were directly interviewed. These 405 interviewees were also asked to prospectively complete a questionnaire for each patient death occurring in the six months subsequent to their interview. Response rate was 80% and yielded questionnaire data on 2257 deaths. Legal immunity was granted by the Minister of Justice to all respondents in these studies. Because this is the only published prevalence survey with an acceptable response rate, findings will be discussed in detail. Euthanasia, defined as the "prescription, supply, or administration of drugs with the explicit intention of shortening life to include euthanasia at the patient's request, assisted suicide, and life terminating acts without explicit and persistent request," occurred in a combined average of 2.9% ($n = 3735$) of all 128,786 deaths occurring in 1990 from all three studies. Assisted suicide accounted for an average of 0.3% ($n = 386$) of deaths, voluntary euthanasia 1.8% ($n = 2318$), and what appeared to qualify as involuntary euthanasia in 0.8% ($n = 1030$). Detailed information on these cases of administration of lethal drugs without explicit and persistent request from patients suggested that in more than half the decision had been previously discussed when the patient was competent, that death was imminent and suffering apparent, and that consultation with family, nurses, and other medical staff nearly always preceded the decision.

Among the interviewed physicians, 54% admitted to practicing voluntary euthanasia or assisted suicide, 34% had never participated but could "conceive of situations in which they would be prepared to do so," and 12% "could not conceive of any situation" in which they would be prepared to assist a patient to die. Patients' reasons for requesting euthanasia included loss of dignity (57%), pain (46%), "unworthy dying" (46%), dependency on others (33%), and "tiredness of life" (23%).

Over 25,000 persons annually request assurance from their doctors that they will receive assistance in dying if needed; 9000 per year actually request euthanasia and fewer than a third of these requests are honored. This survey suggest that the majority of both the public and the medical profession in The Netherlands have accepted the practice of voluntary euthanasia under certain circumstances and appear to understand it as part of a continuum of medical decisions at the end life, including withholding or withdrawing life sustaining treatment, double effect administration of pain relieving medications, and physician-assisted death or voluntary euthanasia. The applicability of these data (from a homogeneous middle-class nation with guaranteed access to health care for all) to the United States is limited, but it provides the first quantitative and descriptive data about a practice now undergoing rigorous public debate here.

A 1991 survey of 1000 internists conducted by the American Society of Internal Medicine,[46] yielded a 40% response rate and found that 25% of respondents had been asked for assisted death by a terminally ill patient, and that 20% of respondents acknowledged "having taken a deliberate action that would directly cause a patient's death." Although two-thirds of respondents said they would vote against legalization of euthanasia, the same majority felt that "suicide was a moral option for someone in great pain." An informal telephone survey of 300 physicians (100 internists, 100 family practitioners, and 100 psychiatrists) conducted by the American Board of Family Practice[47] found that 89–93% agreed that "terminally ill patients have the right to choose to die" and

77–86% believed "patients have the right to choose to die if they have an illness that permanently impairs their quality of life." In contrast, 87% of the internists said there were "no circumstances under which they would administer a lethal dose of a drug," and 78% said it was "unethical for a physician to administer a fatal drug dose." Poor response rate, low numbers, unclear sampling methods, inadequate protection of anonymity, and confusing wording limit the interpretation of these data.

Preliminary written and anonymous survey data of 129 New York City area physicians of various specialties[48] revealed that 85% felt that a terminally ill patient may be capable of a rational decision to commit suicide; 30% felt it should be legal for a doctor to give a lethal injection, and 27% would be willing to do so; 41% felt it should be legal for a doctor to prescribe a lethal dose and 43% would be willing to do so. Eighty-three percent felt they had not had sufficient training in pain management and 94% agreed that doctors need more training in pain management, terminal care, treatment of depression, and hospice options. This survey is limited by its nonrandom and unspecified sampling frame.

A similar questionnaire survey[49] mailed to all 1381 American Board of Internal Medicine-certified internist-geriatricians resulted in a 52.6% response rate to questions about the widely publicized Janet Adkins/Jack Kevorkian case of physician-assisted suicide in a patient with dementia. Sixty-six percent felt that Dr. Kevorkian's actions were unjustified, but 49% felt that Mrs. Adkins' decision to commit suicide was not morally wrong. If responding geriatricians themselves were diagnosed with dementia, 41% would consider suicide a possible option. Fifty-seven percent opposed easing restrictions on physician-assisted suicide, whereas 26% were in favor. If it were legal, 21% would consider assisting in the suicide of a competent dementia patient, whereas 66% would not. This study documents an interesting disparity between respondents' sympathy with the idea that mildly demented persons (including themselves) may be justified in considering suicide vs their rejection of any medical professional participation in such an act. This survey is limited by its specific focus on Dr. Kevorkian, a highly controversial figure whose actions may have elicited more negative physician response on the issue of physician-assisted suicide. Similarly, it addresses the particularly complex problem of whether a patient with dementia could ever be judged competent to make a decision to commit suicide, and thus may not reflect respondents' attitudes toward physician-assisted suicide for patients of intact mental capacity.

Conclusions

These physician survey data are difficult to compare directly because of different study designs, including variable wording and definition of questions, geographic locations, year of conduct, and conduct by advocacy organizations vs more objective sources. Further, although there have been a number of efforts to survey physicians on the subject of physician aid in dying, the questionable validity of the data (owing primarily to very low response rates and the likely unrepresentative nature of those who do respond), and the lack of questions on certain key areas (e.g., how a decision to assist a patient to die is made, the roles of other doctors, nurses, social workers, and family members in the decision-making process) indicate a scarcity of the type of data needed to inform debate on physician aid in dying. In general, previous work has focused primarily on doctors' attitudes toward physician-assisted dying rather than on actual physician actions and the characteristics of those actions. Finally, the intensity of recent public debate on these issues may have led to a change in attitudes (and perhaps in practices as well) of physicians, such that prior work is no longer applicable to current clinical practice.

In spite of the problems with previous survey work, their results do suggest that some fraction of physicians and the general public are in favor of legalizing physician-assisted dying under defined circumstances. Of equal interest is the fact that a significant minority of doctors admitted to assisting their patients to die, in spite of the clear illegality of such actions. This finding suggests that recent highly publicized cases do not represent isolated occurrences, and raise concern that inappropriate, or frankly involuntary instances of physician-assisted dying could be occurring with measurable frequency in this country, as seems to be the case currently in The Netherlands. This finding, if real, should inform the debate on physician-assisted dying and should influence both physician education and public policy efforts with respect to end of life decisions. A more definitive study of actual US physician practices is, however, prerequisite to any change in recommended policy, educational objectives, or practice guidelines in response to these data.

References

[1] A right to die: Debate intensifies over euthanasia and the doctor's role. (1991) *Am. Med. News* **12:** Jan. 7.
[2] Hechinger, F. (1991) They tortured my mother. *New York Times Editorial Notebook,* Jan. 2.
[3] Wariness is replacing trust between healer and patient. (1990) *New York Times* A1, Feb. 20.
[4] Jecker, N. S. (1991) Knowing when to stop: The limits of medicine. *Hastings Centers Report,* May–June 5–8.
[5] Patients don't always respond and get better. (1991) *New York Times Letters to the Editor,* Feb. 13.
[6] Cassel, C. K., Meier, D. E. (1990) Morals and moralism in the debates on euthanasia and assisted suicide. *New Engl. J. Med.* **323,** 750–752.
[7] Annas, G. J. (1990) Nancy Cruzan and the right to die. *New Engl. J. Med.* **323,** 670–673.
[8] In re Quinlan, 70 NJ, 10, 41, 355 A.2d 642, 664, cert. denied, 429 US 922 (1976).
[9] Lo, B., Rouse, F., Dornbrand, L. (1990) Family decision making on trial: Who decides for incompetent patients? *New Engl. J. Med.* **322,** 1228–1232.

[10]Hilfiker, D. (1983) Allowing the debilitated to die. *New Engl. J. Med.* **308,** 716–719.

[11]Dworkin, R. (1991) The right to death. *New York Review of Books,* Jan. 31, pp. 1–17.

[12]Veatch, R. M. (1972) Choosing not to prolong dying. *Med. Dimensions* Dec., 8–40.

[13]Gillick, M. R. (1988) Limiting medical care: Physicians' beliefs, physicians' behavior. *JAGS* **36,** 747–752.

[14]Emanuel, L. L., Barry, M. J., Stoeckle, J. D., Ettelson, L. M, Emanuel, E. J. (1991) Advance directives for medical care: A case for greater use. *New Engl. J. Med.* **324,** 889–895.

[15]Annas, G. J. (1991) The health care proxy and the living will. *New Engl. J. Med.* **324,** 1210–1213.

[16]Angell, M. (1988) Euthanasia. *New Engl. J. Med.* **319,** 1348–1350.

[17]Glover, J. (1977) *Causing Death and Saving Lives.* Penguin Books, New York, pp. 182–189.

[18]Quill, T. E. (1991) Death and dignity: A case of individualized decision making. *New Engl. J. Med.* **324,** 691–694.

[19]Quindlen, A. (1990) Seeking a sense of control. *New York Times,* Editorial, Dec. 9.

[20]Vaux, K. L. (1988) Debbie's dying: Mercy killing and the good death. *JAMA* **259,** 2140, 2141.

[21]Rachels, J. (1975) Active and passive euthanasia. *New Engl. J. Med.* **292,** 78–80.

[22]Cutter, M. A. G. (1991) Euthanasia: Reassessing the boundaries. *J. NIH Res.* **3,** 59–61.

[23]Jury declines to indict a doctor who said he aided in a suicide. (1991) *New York Times* A1, July 27.

[24]Gest, T. (1990) Changing the rules on dying. *U.S. News and World Report,* July 9, 22–24.

[25]Giving death a hand: Rending issue. (1990) *New York Times* A6, June 14.

[26]The Hemlock Society. (1990) 1990 Roper Poll on physician aid-in-dying, allowing Nancy Cruzan to die, and physicians obeying the living will. The Roper Organization, New York, April 24,25.

[27]The Hemlock Society. (1991) 1991 Roper Poll of the West Coast on euthanasia. The Roper Organization, New York, May.

[28]Knox, R. A. (1991) Poll: Americans favor mercy killing. *Boston Globe* Sunday, Nov. 3, A1.

[29]Reidinger P. (1988) Lawpoll. *ABA Journal*, June 1, 20.
[30]Webster's Third New International Dictionary (Unabridged). (1961) G. C. Merriam, Springfield, MA.
[31]Reichel, W., Dyck, A. J. (1989) Euthanasia: A contemporary moral quandary. *Lancet* **ii**, 1321–1323.
[32]Singer, P. A., Siegler, M. (1990) Euthanasia—A critique. *New Engl. J. Med.* **322**, 1881–1883.
[33]Gomez, C. F. (1991) *Regulating Death: Euthanasia and the Case of The Netherlands.* Free Press, New York.
[34]Brahams, D. (1990) Euthanasia in the Netherlands. *Lancet* **335**, 591,592.
[35]Leenan, H. J. J. (1990) Coma Patients in the Netherlands. *BMJ* 300, 69,33.
[36]Van Der Maas, P. J., Van Delden, J. J. M., Pijnenborg, L., Looman, C. N. (1991) Euthanasia and other medical decisions concerning the end of life. *Lancet* **338**, 669–674.
[37]Anonymous. (1988) It's over, Debbie. *JAMA* **259,** 272.
[38]Meier, D. E., Cassel, C. K. (1983) Euthanasia in old age: Case study and ethical analysis. *J. Am. Geriatr. Soc.* **31,** 294–298.
[39]Kuhse, H., Singer, P. (1988) Doctors' practises and attitudes regarding voluntary euthanasia. *Med. J. Australia* **148,** 623–627.
[40]National Opinion Poll Market Research. (1987) Attitudes towards euthanasia among Britain's GPs. NOP Market Research Ltd., London.
[41]The National Hemlock Society. (1988) 1987 Survey of California physicians regarding voluntary active euthanasia for the terminally ill. Feb. 17.
[42]Center for Health Ethics and Policy, University of Colorado. (1988) Withholding and withdrawing life sustaining treatment: A survey of opinions and experiences of Colorado physicians. Graduate School of Public Affairs, Denver, CO, May.
[43]Heilig, S. (1988) The SFMS Euthanasia survey: Results and analyses. *San Francisco Med.* 24–34 (May).
[44]Overmyer, M. (1991) National survey: Physicians' views on the right to die. *Physician's Management* **31,** 40–45.
[45]Hiller, R. J. (1991) Ethics and Hospice Physicians. *Am. J. Hospice Palliative Care,* Jan./Feb., 17–26.
[46]Crosby, C. (1992) Internists grapple with how they should respond to requests for aid-in-dying. *The Internist,* March 10.

[47]Charnow, J. A. (1991) Most internists support patients' right to die, surveys find. *ACP Observer,* June 8.

[48]Devons, C. A. J., Mulvihill, M. (1992) Physician attitudes about physician-assisted death. *The Gerontologist* **32,** p. 70, October special issue.

[49]Watts, D. T., Howell, T., Priefer, B. A. (1992) Geriatricians' attitudes toward assisting suicide of dementia patients. *J. Am. Geriatr. Soc.* **40,** 878–885.

The Nonnecessity of Euthanasia

Gregg A. Kasting

Among the arguments against legalizing voluntary euthanasia and physician-assisted suicide, there is one yet to be scrutinized in the literature. This argument states simply that these acts are not necessary. The reason they are not necessary, it is claimed, is that pain is manageable in "virtually all" or, at least, "most" ⌖ patients with terminal illness.

There are three telling criticisms of the "nonnecessity" argument. First, available data in the medical literature do not support the claim that virtually all cases of severe pain in terminally ill patients can be relieved satisfactorily. On the contrary, recent studies document significantly less than complete success in treating pain in terminally ill patients. Second, when the nonnecessity argument is grounded on the premise that *most* pain in terminal illness is relievable, the most one may conclude is that euthanasia and physician-assisted suicide are unnecessary in those cases where pain and suffering *are* relievable. Any attempt to argue from the premise that most pain is relievable to the conclusion that *all* aid-in-dying is unnecessary will be question-begging. Last, the nonnecessity argument presupposes that severe physical pain is the only cause of extreme suffering that might justify a request for aid-in-dying. Although this may be true in a majority of cases, other components of terminal suffering are completely ignored by this argument.

Examples of the Nonnecessity Argument

In a letter published in the *New England Journal of Medicine* in 1975, Robert Sade and Anne Redfern responded vigorously to James Rachels' paper, "Active and Passive Euthanasia."[1] They wrote, "the process of 'being allowed to die' need never be painful, though Rachels implies that it often is . . . . Since the physician always has the means to relieve suffering, active termination of life lies outside his moral obligation to his patient."[2]

Sixteen years later, in another letter to the same journal, Charles Cavagnaro and David O'Brien again made the nonnecessity argument, although somewhat less absolutely. These authors, replying to another piece on assisted death,[3] stated the argument this way: "Our present ability to manage most forms of pain associated with terminal illness effectively weakens most of the arguments favoring euthanasia. So why do Cassel and Meier ask us to rejoin the discussion?"[4]

In another recent piece, Kevin O'Rourke strongly rejected the argument that physician-assisted death is morally justified to relieve "unbearable pain."[5] He took the nonnecessity argument a step further in blaming the physician for the predicament of a patient who requests aid-in-dying:

> (T)he ability to limit and remove pain is within the armamentarium of the physician in almost every case. The situation in which pain cannot be controlled probably reflects more upon the expertise of the physician than upon the severity of the disease.

Be it in the strong ("virtually all pain is relievable") or the weak form ("most pain in terminal illness is relievable"), some version of the nonnecessity argument appears regularly in pieces opposing voluntary assisted death. Citations of clinical data that might substantiate either of these two claims are rarely given, however. In fact, if one follows the trail of citations that leads through successive papers, it is quite apparent that this central question— how relievable is pain in terminal illness?—has escaped careful scrutiny throughout the literature on euthanasia. The following are typical examples.

In their 1990 paper, "Indirectly Intended Life-Shortening Analgesia: Clarifying the Principles," Robert Barry and James Maher asserted the strong form of the nonnecessity argument as follows:

> Analgesic medicine has changed so radically that the President's Commission for the Study of Ethical Problems in Medicine and Biomedical and Behavioral Research claimed that recent developments have allowed pain to be reduced to an acceptable level in virtually all cases. It is true that in areas where medicine is less advanced this might not be the case, but where more advanced medicine is being practiced, the reasons given for lethal overdosing as medically necessary are usually now considered to be invalid.
>
> Joseph Fletcher has written that the best possible news advocates of euthanasia could hope to have would be that medical advances have made euthanasia no longer necessary. With few exceptions, it can now be said that medicine has fulfilled this dream within the past ten years.[6]

The President's Commission did, indeed, make the claim as Barry and Maher cited. Its 1983 report stated,

> Medical management of symptoms has recently demonstrated that no patient should have to be terrified of physical pain; in fact, presently available drugs and techniques allow pain to be reduced to a level acceptable to virtually every patient, usually without unacceptable sedation.[7]

In support of this rather sweeping statement one might think that the commission would have provided a review of the relevant literature. This was not the case, as it cited just two texts on the management of terminally ill patients.[8]

From the first of these, the commission referenced a chapter on pain relief authored by Robert Twycross.[9] In it, the only data presented on the relievability of cancer pain was from a study by

Parkes, published in 1977.[10] Contrary to the commission's strong statement, Parkes' data did not show pain relief in virtually all patients: 36% of patients studied had severe, unrelieved pain in the preterminal phase of their illness, and 8% had pain of this magnitude in the terminal phase.

The second text that the commission cited, a collection of papers published in 1973, included clinical data on the use of a variety of analgesics and adjuvant medications in terminal care.[11] None of these reports, including one by Twycross on the use of heroin, gave any data to show what percentage of patients have severe pain despite therapy. Quite contrary to the commission's very optimistic assessment, one author in the collection stated the following:

> This review of the use of drugs for the dying patient leaves one with a sense of disquiet and concern. Almost always, drugs are an integral part of man's encounter with death and yet the quality of evidence that we have about the value of psychopharmacologic drugs in the treatment of the dying patient is poor, indeed. . . . Hopefully, in the coming decade, we can expect that research on the place of drugs for the dying patient will be substantially upgraded.[12]

A second example showing a lack of critical examination of this question appears in the arguments of Peter Singer and Mark Siegler. In their paper, "Euthanasia—A Critique," the authors presented the weak form of the nonnecessity argument as follows:

> Most physical pain can be relieved with the appropriate use of analgesic agents. Unfortunately, despite widespread agreement that dying patients must be provided with necessary analgesia, physicians continue to underuse analgesia in the care of dying patients because of concern about depressing respiratory drive or creating addiction. Such situations demand better management of pain, not euthanasia. . . .
>
> These days, competent patients may freely exercise their right to choose or refuse life-sustaining treatment; to carry out their preferences, they do not require the option of euthanasia.[13]

Singer and Siegler provided two references for these statements. One, a 1982 *New England Journal of Medicine* editorial by Marcia Angell decrying the undertreatment of pain, stated the following: "It is generally agreed that most pain, no matter how severe, can be effectively relieved by narcotic analgesics."[14] In support of this statement, Angell cited three references; all were descriptive papers— not studies of clinical trials—that discussed pharmacologic methods of pain relief in terminal care.[15] Only two of these articles commented on how relievable cancer pain is, and neither of them provided or cited any data to substantiate the individual authors' claims.

In the first of these two, the authors stated "with care and patience, the physician can render practically any cancer patient pain-free." In their conclusion, however, they were somewhat more cautious saying, "most patients can be made relatively pain-free provided their pain is managed with patience in an individualized fashion."[16] The author of the second paper, Twycross, stated in 1975,

> To most people pain and incurable cancer are inextricably intertwined. In fact, as many as half of all patients with terminal cancer have no pain or negligible discomfort at most. Forty percent experience severe pain and the remaining 10% suffer less intense pain. Furthermore, it is theoretically possible to relieve pain in every case.[17]

It is not remarkable that Twycross, long affiliated with St. Christopher's Hospice in London, gave no reference concerning the prevalence of pain among terminal cancer patients. It is somewhat remarkable, though, that he gave no further discussion or data in support of the very strong statement that, theoretically, pain is relievable in every case of terminal cancer.

Returning to Singer and Siegler, the second reference they provided was a review article on techniques for the treatment of cancer pain. Although Kathleen Foley, its author, did not explicate how much might be a result of undertreatment or refractoriness, she wrote, "it is postulated that 25% of all patients with cancer throughout the world die without relief from severe pain."[18]

Having examined these citations that give little support to the central claim of the nonnecessity argument, several points seem noteworthy. First, if it is the case that all cancer patients who die in severe pain do so because their pain is undertreated for one reason or another, then certainly euthanasia as a release from unrelieved pain is unnecessary. Do any of the authors who make the nonnecessity argument know of data that show this is the case? On the other hand, if a full 25% of cancer patients die with severe pain that is refractory to treatment, how significant or forceful as a reason for disallowing assisted death is the fact that merely "most" physical pain can be relieved? Last, if only *some* of the patients who die with severe pain do so because of undertreatment, what portion of this 25% is it?

For the nonnecessity argument to carry any force whatever, it must be accompanied by some data on the successfulness of the techniques used to relieve pain in terminal illness. What is troubling about the argument is that little or no clinical data are referenced each time the argument appears. When one follows the trail of citations from one paper to the next, one finds only a series of unsubstantiated and amplified claims.

All of this is not to question whether presently available analgesics, anesthetic techniques, and the efforts of those providing terminal care are often efficacious in relieving terminal pain; certainly they are. The question is, to what degree are they efficacious? This is obviously an empirical question, however, and it should be answered by an appeal to solid clinical data, not conjecture.[19]

A Brief Review
of Pain Relief in Terminal Care

The nonnecessity argument clearly presupposes that the deliverance from physical pain is the only reason patients might request physician-assisted death. As we will discuss later, there are many other facets of suffering at the end of life, even though physical pain is one of the most common.

For the moment, let's assume that physical pain is the only cause of suffering that would lead patients to request euthanasia or aid-in-dying. Again, the question we must answer is the following: According to clinical studies, how successful is our health care system at relieving pain in terminal illness? Is it anywhere close to "virtually all"? If not, what is a good estimate of the ubiquitous quantifier—most—that authors use when they state most pain in terminal illness is relievable?

Before searching the medical literature for an answer to this question, some background points bear stating. First, although the two are frequently used interchangeably, "terminal illness" is not synonymous with "terminal cancer." Obviously the latter is a subset of the former, since there are terminal illnesses apart from malignancies.

Second, pain is indeed a common problem with terminal cancer. The overall prevalence of pain among all cancer patients is estimated to be 68%, and the prevalence of severe pain in patients with advanced cancer is put at 75–80%.[20]

Third, for over a decade it has been reported that physicians on the whole do a poor job of treating pain in terminal illness.[21] Bonica reviewed twelve studies, dating from 1959 to 1985, that showed rates of unrelieved pain from 32–80%. He commented as follows:

> (W)hy is cancer pain inadequately relieved? Serious consideration of this important question by this writer in the course of the past three decades has repeatedly suggested that it is due to an inadequate appreciation or outright neglect of the problem of *pain* (in contrast to the problem of cancer) by oncologists, medical educators, investigators, research institution, and national and international cancer agencies.[22]

In reviewing these twelve studies, Bonica implied that the data do not reflect the pain relief that could be attained by empathetic, diligent, and skilled caregivers. If we grant this, then clearly we should not seek from the clinical literature simply what percentage of *all* patients with terminal illnesses obtain acceptable pain relief. To answer the question at hand we should seek instead what percentage of terminal patients obtain relief under the *best* of circumstances, that is, when treated by experienced and knowledgeable clinicians who are dedicated to terminal care.

Last, we should examine the prevalence of pain relief/nonrelief not in patients at *all* stages of their illnesses, but those who are indeed *terminal*. Obviously, for the nonnecessity argument to carry any weight it must hold up in the end-of-life settings where patients are the most likely to request aid-in-dying.

As one might surmise from the examples in the preceding section, there are indeed very little good data that show the degree to which pain relief is achievable in terminal illness. There are, however, at least four studies from the past several years that cast doubt on the claims made above.

The first and smallest study of pain relief in terminal illness that will be reviewed here came from the Edmonton General Hospital and Cross Cancer Institute, Edmonton, Alberta, Canada. The authors examined the cases of 48 patients who were treated by their palliative care team in 1984, and 50 such patients who were treated in 1987. They reported that

> Twenty-nine percent and 37% of the patients died in poor pain control under our care in 1984 and 1987, respectively. However, these percentages are biased because 50 and 60% of the patients referred to the PCT (palliative care team) during 1984 and 1987, respectively, were discharged to other hospitals. The greater majority of these patients were in good pain control . . . .

> Our results suggest that a significant portion of patients with advanced cancer have pain syndromes that do not respond to recommended treatments. Therefore, education alone will not result in the elimination of the problem of cancer pain. More research is necessary in order to better define those intractable pain syndromes and to develop adequate treatments for them.[23]

If we grant that *all* of the team's patients who were referred
to other institutions were in states of good pain control, then the
percentage of original patients who were left with "poor" pain
control can be calculated. Assuming that all—not just a majority—of the referred patients had good pain relief, it turns out that
the portion of the initial group who experienced inadequate pain
relief was 15% in each year of study.

Admittedly, this is a rather small study, and without knowing how representative their sample is of all terminally ill patients
one should not draw broad conclusions from these data alone.
The statements by this group of terminal care specialists, however, make one question the basis of the nonnecessity argument.

Further telling data and comments come from a study conducted at the National Cancer Institute in Milan, published in
1990.[24] The authors looked at the courses of 115 cancer patients
whose diseases had stopped responding to anticancer treatments.
They reported that 74% of these patients required "strong" narcotics, and that a full 35% reported uncontrolled pain during their
palliative care. Regarding the failure to relieve pain adequately in
such a large number of their patients, the Milan authors wrote,

> In accord with data in the literature, we found intense pain
> to be the most frequently occurring symptom. During the
> week of observation, the intake of only analgesic drugs
> reduced the number of patients with uncontrollable pain by
> two-thirds, indicating that pain control is not always pos-
> sible in some patients.

> Present methods of data analysis, which adopt the mean or
> median of pain control, fail to show the number of patients
> with uncontrolled pain. Our data show that full control of
> this symptom is still quite hard to achieve and maintain during
> the entire period of treatment despite progress made in therapy
> over the last few years.[25]

Another recent study documenting the success rate of cancer
pain specialists was done by Dwight Moulin and Kathleen Foley at

Memorial Sloan-Kettering Cancer Center.[26] These authors reported on 165 patients who were referred by their attending physicians for "diagnostic or therapeutic pain problems" to Sloan-Kettering's multidisciplinary group of pain consultants.

Like the teams at the Edmonton General Hospital and the National Cancer Institute in Milan, this group was unable to demonstrate the purported mastery over cancer pain that some authors claim exists. Moulin and Foley reported, "overall, 59 patients (38%) had excellent pain relief, 55 (35%) had moderate relief, and 43 (27%) had a poor outcome." For comparison, these authors also cited Parkes' data from St. Christopher's Hospice in London where 36% of the patients had severe and unrelieved pain in the preterminal phase of their illness, and 8% had pain of the same magnitude during their terminal period.[27] The Sloan-Kettering authors commented,

> It is clear from these studies that even when analgesic agents were used appropriately and other techniques were applied, one-quarter to one-third of patients had unacceptable pain relief or side effects that seriously interfered with the quality of life. Recent advances, including the use of continuous subcutaneous narcotic infusion and patient-controlled analgesia, may reduce this figure to 10 percent to 20 percent.[28]

It should be emphasized that the patients in this study were selected not from all cancer patients but from that subset of patients whose pain was a diagnostic or therapeutic problem. Again, without knowing what percentage of all cancer patients this group of 165 difficult cases comprises, it is not clear from these data alone what percentage of all terminally ill patients have unrelieved pain.

The largest collection of data that might answer this question comes from the National Hospice Study, published in 1986. This study detailed the experiences of 1754 patients with terminal cancer.[29] Of this number, 833 patients were cared for by home-based hospices, 624 were cared for by hospital-based hospices, and 297 patients were cared for in conventional hospital settings.

Among the hospice patients who were contacted three to eight days before death, 19% of the home-based patients and 15% of the hospital-based hospice patients described themselves as being in severe pain. Of 221 hospice patients contacted two days before death, 21% said they were experiencing severe pain. (For comparison, among conventional care patients 28% complained of severe pain three to eight days before death, and 41% described this level of pain two days before death.) In their discussion the authors of this study stated,

> As the largest study of its type to date, the findings transcend many limitations of earlier studies. Data presented confirm that terminal cancer does not invariably mean an agonizingly painful death. Twenty to thirty percent of patients are pain free during their final weeks of life, and this proportion remains fairly stable as death approaches. However, the prevalence of persistent or severe pain does increase as death approaches.[30]

It is rather obvious that the statement, "terminal cancer does not invariably mean an agonizingly painful death," is a far cry from the claim that pain can be reduced to an acceptable level in all cases. If we look only at the results from the hospital-based hospice patients—who received the most intensive and expert attention of the three groups and still had unrelieved, severe pain in 15%—we see, once again, no support for the central premise of the strong nonnecessity argument.

These four recent studies describe the success of terminal care specialists in Canada, Italy, and the United States. At the time three of these reports were published, some seven years after the President's Commission's optimistic report in 1983, palliative care providers failed to relieve pain in 15–35% of the patients referred to them. Although this may not be a precise answer to the question we posed above, these data do make one point very clear. The claim that "all" or "virtually all" pain is relievable in terminal illness does not appear to be true at the present time, even under the best of circumstances.

The Weak Form
of the Nonnecessity Argument

Since it is questionable, at best, that "virtually all" pain in terminal illness is relievable, what proponents of the nonnecessity argument are left with as an acceptable premise is the claim that "most" pain in such situations is relievable. Will this sustain the argument, however?

If opponents of euthanasia and assisted suicide concede that some percentage of terminal patients may suffer severe, intractable pain, and still wish to press the nonnecessity argument, then logically euthanasia is unnecessary only in those cases where pain or suffering is relievable. Certainly, however, few if any who support the practice of physician-assisted death on the basis of beneficence would argue against denying the request of a patient whose suffering is tractable. That a patient's suffering must be unrelievable is one of the criteria for granting a request for euthanasia in the Netherlands.[31] Other authors proposing the legalization of physician-assisted suicide in the United States have outlined procedures that stipulate this as well.[32]

Still, many authors make the nonnecessity argument on the claim that "most" pain in terminal illness is relievable. One wonders how these authors can argue that because only a numerical minority of cancer patients suffer severe and intractable pain, it is "unnecessary" to legalize euthanasia and physician-assisted suicide. One imagines they reason implicitly that the problem is not large enough to "necessitate" changes in medical practice, legal statutes, and social mores. If this is indeed their line of reasoning, several things must be said in response.

First, were we to accept this reasoning without any additional considerations, the argument would beg outright the question of whether physician-assisted death is permissible for those patients who experience intractable suffering. If the reasons in favor of assisting the death of a terminally ill patient are otherwise sound, then the act should not be proscribed based on how many other people are, or are not, in a similar situation.

Responding to this very point, some opponents of physician-assisted death might grant that a single act of euthanasia, performed at the request of a rational patient with severe and uncontrolled suffering, might be morally permissible. They would then lay out the "slippery slope" argument that legalized voluntary euthanasia as a routine practice will lead to a host of grevious consequences— mainly involuntary euthanasia—that on balance will far outweigh any good that might result. With many authors now examining the Dutch experience, a benign outcome of "legally tolerated" (now legalized) and socially accepted assisted death in the Netherlands is being questioned. A large number claim that the country is witnessing the serious problems and widespread abuses opponents of legalized euthanasia feared.[33] Whether procedural safeguards can be found to avoid involuntary euthanasia is beyond the scope of this discussion. However, this much can be said: The slippery slope argument does not arise from the premise that assisted death is unnecessary, nor does it demonstrate that these acts are unnecessary.

Second, this implicit argument fails to recognize that given the prevalence of diseases that cause significant terminal suffering, "small" percentages turn out to be large actual numbers of people. Along this line, one also wonders whether these authors do not substantially underestimate the magnitude of the situation. Consider that in 1991 an estimated 516,170 Americans died from cancer.[34] If 5% of these people died with unrelieved and severe pain, the total number of people dying under such circumstances would equal 25,809. Previewing the next section, consider that if 5% of the 908,780 people who died from major cardiovascular diseases,[35] 5% of the 88,690 who died from chronic obstructive pulmonary diseases,[36] and 5% of the 73,980 who died from pneumonia in the United States in 1991 experienced severe and intractable dyspnea (shortness of breath or difficulty in breathing) in their last week of life, it would mean that 53,573 more Americans died suffering. The "small" portion of 5% of those who suffer unacceptably from pain and dyspnea within these groups would thus represent over 79,000 patients dying in one year in the United States.

Now recall the comments of the four groups of palliative care specialists above on their estimates of patients who suffer severe, unrelieved terminal pain despite their care. Consider further that 24% of the patients in the National Hospice Study suffered from "horrible" or "severe" dyspnea in their last week of life.[37] Consider even further that this calculation has not included those patients who suffer inexorable deterioration at the end of life from nonmalignancies, such as AIDS and progressive neurological diseases, among others. Given the sizable number of all patients who experience unremitting suffering *(see below)*, the simple fact that "most" terminal pain or suffering is relievable does not allow one to conclude that the legalization of voluntary assisted death is "unnecessary."

Suffering Other Than Pain

The most devastating criticism of the nonnecessity argument is this: Pain is by no means the only facet of suffering that visits dying patients. Other debilitating physical symptoms associated with terminal illness are numerous. Table 1 is a list of symptoms taken from four surveys of terminal patients.[38] Three involved terminal cancer patients, and one reported on patients who suffered from a variety of illnesses.

Most people would agree that the symptoms listed in Table 1 produce significant suffering. The primary goal in terminal care is to relieve or palliate these problems with medications, physical aids, surgical procedures, and attentive nursing care. All too often, however, these problems are not alleviated; the percentage of patients reporting uncontrolled symptoms in two of these studies reflect this.[39] Furthermore, patients are very rarely afflicted by just one of these symptoms, but rather are beset by a constellation of them at the same time.

Perhaps the most distressing physiologic experience seldom mentioned in this debate is dyspnea, or shortness of breath. This

Table 1
Common Symptoms of Terminally Ill Patients

Symptom	Percent of total
Weakness	39–91
Pain	49–62
Anorexia	8–76
Immobility	41[a]
Constipation and bowel disorders other than incontinence	4–51
Dyspnea	17–51
Urinary incontinence	35[a]
Cough	6–45
Nausea/vomiting	9–44
Difficulty in swallowing	4–25
Insomnia	7–24
Confusion	9–24[b]
Pressure sores	14[a]
Fecal incontinence	13[a]
Offensive odors	5[a]

[a]Symptom listed only in Wilkes (1984).
[b]Symptom not listed in Saunders and Baines (1989).

symptom is, in fact, quite prevalent in terminal illness. In the National Hospice Study, 70% of all terminal cancer patients reported some degree of dyspnea during the last six weeks of life. As mentioned above, of 878 patients who were interviewed during their last week of life, 24% described their degree of air hunger as "horrible" or "severe."

The disease mechanisms by which cancer produces dyspnea are many.[40] In addition to these, there are the well known, non-cancer related processes that cause air hunger in patients both with and without malignancies: congestive heart failure, chronic obstructive pulmonary disease, pneumonia, and pulmonary emboli.

The obvious question is, can dyspnea be relieved? In many cases, underlying pathological processes can be treated. It is an unfortunate reality in hospitals everywhere, however, that when many of these mechanisms are present in an advanced stage, respiratory failure is ameliorable only with endotracheal intubation and ventilatory support. But for several obvious reasons, respirators are rarely used for terminally ill patients. In those instances, the distress of dyspnea may continue until the patient develops consciousness-altering hypoxemia (low levels of oxygen in the blood) and hypercapnia (high levels of carbon dioxide in the bloodstream).

Some authors report that air hunger in terminal illness can be well managed with sedatives or narcotics.[41] Data in support of this, however, is quite scarce—more scarce than clinical data on the relievability of pain in terminal illness. There are some that show that dyspnea is not easily treated,[42] and other data to indicate that "relief" comes at the cost of somnolence and even death.[43]

It is obvious, however, that the argument that voluntary assisted death is unnecessary because very efficacious analgesics and anesthetic techniques are presently available wrongly focuses on physical pain alone. Given all of the other elements of suffering that accompany the end of life, this reasoning is scarcely compelling. For the argument to remain viable, its proponents would have to show that all significant elements of terminal suffering are fully tractable.

Furthermore, those who wish to make the nonnecessity argument also must demonstrate that the patients' symptoms truly can be relieved, and are not merely blotted out by drug-induced somnolence or coma. Some authors have announced that by using the World Health Organization's guidelines for the treatment of cancer pain, only 3% of patients experienced severe or very severe pain at the time of death.[44] What the authors mentioned, but did not highlight, was that at the time of their death, 20% of the patients were unable or unwilling to rate their pain—yet all were included in the 97% who were said not to have severe pain. It is certainly true that a fully anesthetized patient is unable to feel or report pain or dyspnea, but is it right to call this relief? One imagines that to "relieve" a symptom—be it pain, dyspnea, or whatever—the patient's quality of life must be improved, not merely obliterated.

Conclusion

All agree that today's medical technology is able to palliate cancer pain in a majority of cases. Recent studies indicate that a sizable number of cancer patients receive inadequate relief of suffering despite the best efforts of physicians, nurses, and pharmacists who are dedicated to alleviating their pain. Even so, some ethicists and physicians glibly assert that all pain in terminal illness is completely mitigable.

Furthermore, it is possible that the complete relievability of cancer pain, without excessive or equally troubling side effects, may never become a reality despite advances in technology and physician education. It is also possible that despite progress in other areas of palliative care there will continue to be some terminally ill patients who experience elements of unrelieved and intolerable suffering. So long as this remains the case, the issue of physician-assisted death will be discussed.

The argument that assisted death is unnecessary cannot be made on the claim that pain in terminal illness is amenable in full or in part to presently available analgesics and anesthetic techniques. As we have seen, the empirical claim that all terminal pain is relievable is not supported by the medical literature. Also, to push a weaker form of this argument (based on the premise that most pain is relievable) obviously begs the question in those cases where pain is not malleable. This weak version of the argument may also beg the question by implicitly and wrongly minimizing the magnitude of terminal suffering. Finally, this reasoning incorrectly assumes that pain is the only significant cause of suffering and therefore represents the only reason patients might request physician-assisted death. In doing so, it ignores the many other elements of physical suffering, particularly dyspnea. For these reasons, in the ongoing debate surrounding voluntary physician-assisted death the nonnecessity argument can be rejected in full.

Medical science certainly will continue its quest to relieve pain and other forms of suffering. All parties on both sides of this issue pray that health care providers will always give great attention to the care of the dying. A rejection of the nonnecessity argument, however, is grounded on the realities of terminal illness. Anyone who has seen the difficult death of a patient or loved one who suffered despite the best efforts of competent and sympathetic caregivers knows this. One may still argue that voluntary euthanasia and assisted suicide should remain impermissible for other reasons, but the argument that these acts are unnecessary simply does not hold up.

Notes and References

[1]J. Rachels (1975) Active and passive euthanasia, *New Engl. J. Med.* **292:2,** 78–80.

[2]R. M. Sade and A. B. Redfern (1975) Euthanasia (letter), *New Engl. J. Med.* **292:16,** 863,864.

[3]C. K. Cassel and D. E. Meier (1990) Morals and moralism in the debate over euthanasia and assisted suicide, *New Engl. J. Med.* **323:11,** 750-752.

[4]C. E. Cavagnaro and D. L. O'Brien (1991) Assisted suicide and euthanasia (letter), *New Engl. J. Med.* **324:20,** 1435, 1436.

[5]K. O'Rourke (1991) Assisted suicide: An evaluation, *J. Pain Symptom Manage.* **6:5,** 317–324.

[6]R. Barry and J. E. Maher (1990) Indirectly intended life-shortening analgesia: Clarifying the principles, *Issues Law Med.* **6:2,** 123, 124.

[7]President's Commission for the Study of Ethical Problems in Medicine and Biomedical and Behavioral Research (1983) *Deciding to Forego Life-Sustaining Treatment.* US Government Printing Office, Washington, DC, pp. 50,51.

[8]C. M. Saunders, ed. (1978) *The Management of Terminal Disease,* Edward Arnold, London; I. K. Goldberg, S. Malitz, and A. H. Kutscher, eds.

(1973) *Psychopharmacological Agents for the Terminally Ill and Bereaved,* Columbia University Press, New York.

[9]R. G. Twycross (1978) Relief of pain, in *The Management of Terminal Disease* (C. M. Saunders, ed.), Edward Arnold, London, pp. 65–92.

[10]C. M. Parkes (1977) Evaluation of family care in terminal illness, in *The Family and Death* (E. R. Pritchard, J. Collard, B. A. Orcutt, A. H. Kutscher, I. Seeland, and N. Lefkowicz, eds.), Columbia University Press, New York.

[11]Goldberg, Malitz, and Kutscher (1973).

[12]G. L. Klerman (1973) Drugs and the dying patient, in *Psychopharmacological Agents for the Terminally Ill and Bereaved* (I. K. Goldberg, S. Malitz, and A. H. Kutscher AH, eds.), Columbia University Press, New York, p. 22.

[13]P. A. Singer and M. Siegler (1990) Euthanasia—A critique, *New Engl. J. Med.* **322:26,** 1881, 1882.

[14]M. Angell (1982) The quality of mercy, *New Engl. J. Med.* **306:2,** 98.

[15]W. T. Beaver (1980) Management of cancer pain with parenteral medication, *JAMA* **244:23,** 2653–2657; D. S. Shimm , G. L. Logue, A. A. Maltbie, and S. Dugan (1979) Medical management of chronic cancer pain, *JAMA* **241:22,** 2408–2412; R. G. Twycross (1975) Relief of terminal pain, *Br. Med. J.* **4,** 212–214.

[16]Shimm et al. (1979), pp. 2408, 2412.

[17]Twycross (1975), p. 212.

[18]K. M. Foley (1985) The treatment of cancer pain, *New Engl. J. Med.* **313:2,** p. 85.

[19]R. L. Daut and C. S. Cleeland (1982) The prevalence and severity of pain in cancer, *Cancer* **50:9,** 1913–1918.

[20]J. J. Bonica (1985) Treatment of cancer pain: Current status and future needs, in *Advances in Pain Research and Therapy* (H. T. Fields, R. Dubner, and F. Cervero, eds.), Raven, New York, vol. 9, 589–616.

[21]J. J. Bonica (1979) Importance of the problem, in *Advances in Pain Research and Therapy* (J. J. Bonica, V. Ventafridda, R. B. Fink, L. E. Jones, and J. D. Loeser, eds.), Raven , New York, vol 2, p. 4. W. T. McGivney and G. M. Crooks, eds. (1984) The care of patients with severe chronic pain in terminal illness, *JAMA* **251:9,** 1182–1188. World Health Organization (1986) *Cancer Pain Relief.* World Health Organization, Geneva.

²²J. J. Bonica (1985), p. 599.

²³E. Bruera, K. MacMillan, J. Hanson, and R. N. MacDonald (1990) Palliative care in a cancer center: Results in 1984 versus 1987, *J. Pain Symptom Manage.* **5:1,** 4, 5.

²⁴V. Ventafridda, F. De Conno, C. Ripamonti, A. Gamba, and M. Tamburini (1990) Quality-of-life assessment during a palliative care programme, *Annals. Oncol.* **1:6,** 415–420.

²⁵*Ibid.,* pp. 417, 418.

²⁶D. E. Moulin and K. M. Foley (1990) A review of a hospital-based pain service, in *Advances in Pain Research and Therapy* (K. Foley et al., eds.), Raven, New York, vol. 16, pp. 422, 425.

²⁷C. M. Parkes (1978) Home or hospital? Terminal care as seen by surviving spouses, *J. Royal Coll. Gen. Pract.* **28,** 19–30.

²⁸Moulin and Foley (1990), p. 425.

²⁹J. N. Morris, V. Mor, R. J. Goldberg, S. Sherwood, D. S. Greer, and J. Hiris (1986) The effect of treatment setting and patient characteristics on pain in terminal cancer patients: A report from the National Hospice Study, *J. Chronic Disease* 39: 1, 27–35.

³⁰Morris et al. (1986), p. 33.

³¹M. A. M. de Wachter (1989) Active Euthanasia in the Netherlands, *JAMA* **262:23,** 3316–3319. H. Rigter, E. Borst-Eilers, H. J. J. Leenen (1988) Euthanasia across the North Sea, *Br. Med. J.* **297,** 1593–1595.

³²T. E. Quill, C. K. Cassel, and D. E. Meier (1992) Care of the hopelessly ill: Proposed clinical criteria for physician-assisted suicide, *New Engl. J. Med.* **327:19,** 1380–1384. H. Brody (1992) Assisted death—A compassionate response to a medical failure, *New Engl. J. Med.* **327:19,** 1384–1388. G. I. Benrubi (1992) Euthanasia— The need for procedural safeguards, *New Engl. J. Med.* **326:3,** 197–199.

³³C. P. Gomez (1991) *Regulating Death: Euthanasia and the Case of the Netherlands,* Free, New York.

³⁴Centers for Disease Control (1992) *Monthly Vital Statistics Report* **41:1,** 16, 17.

³⁵*Ibid.,* p. 16. Major cardiovascular diseases include valvular heart disease, hypertensive heart and renal disease, and ischemic heart disease, all of which are significant causes of or contributors to congestive heart failure and dyspnea.

[36]*Ibid.*, p. 17. Chronic obstructive pulmonary diseases include asthma and emphysema. Deaths from these diseases would be accompanied by dyspnea in an overwhelming majority of patients who were not placed on ventilators, far beyond the 5% used in this calculation.

[37]D. B. Reuben and V. Mor (1986) Dyspnea in terminally ill cancer patients, *Chest* **89:2**, 234–236.

[38]C. M. Saunders and M. Baines (1989) *Living with Dying: The Management of Terminal Disease,* 2nd ed., Oxford University Press, New York, pp. 32,33. E. Wilkes (1984) Dying now, *Lancet* **1,** 950–952. N. Coyle, J. Adelhardt, K. M. Foley, and R. K. Portenoy (1990) Character of terminal illness in the advanced cancer patient: Pain and other symptoms during the last four weeks of life, *J. Pain Symptom Manage.* **5:2,** 83–93. Ventafridda et al. (1990).

[39]Ventafridda et al. (1990) and Wilkes (1984).

[40]Reuben and Mor (1986), p. 235.

[41]L. Heyse-Moore (1984) Control of other syptoms: respiratory symptoms (section), in *The Management of Terminal Malignant Disease,* 2nd ed. (C. M. Saunders, ed.), Edward Arnold, London, pp. 113–119. M. H. Levy and R. B. Catalano (1985) Control of common physical symptoms other than pain in patients with terminal disease, *Semi. Oncol.* **12:4,** 411–430.

[42]I. Higginson and M. McCarthy (1988) Measuring symptoms in terminal cancer: Are pain and dyspnea controlled? *J. Royal Soc. Med.* **82,** 264–267.

[43]M. H. Cohen, A. Johnston-Anderson, S. H. Krasnow, and R. G. Wadleigh (1992) Treatment of intractable dyspnea: Clinical and ethical issues, *Cancer Invest.* **10:4,** 317–321.

[44]S. Grond, D. Zech, S. A. Schug, J. Lynch, and K. A. Lehann (1991) Validation of World Health Organization guidelines for cancer pain relief during the last days and hours of life, *J. Pain Symptom Manage.* **6:7,** 411–422.

The Constitutionality of Elective and Physician-Assisted Death

G. Steven Neeley

"The best form of death is the one we like."
—Seneca, *Ad Lucilium Epistulae Morales*

"Many die too late, and a few die too early. The doctrine still
sounds strange: 'Die at the right time!'"
—Nietzsche, *Thus Spoke Zarathustra*

The present debate over physician-assisted death can be
nothing more than speculative musing unless the patient has a
legal right to command assisted death and the physician is, at
least, legally permitted to assist. Yet the present state of the law
often makes it extremely difficult for the individual to have unfettered control over the act of dying. Twenty-six states and the Com-
monwealth of Puerto Rico currently have statutes that specifically
outlaw assisting suicide.[1] Three additional states would appar-
ently hold one who assisted a suicide guilty of murder as a principle.[2] Several states would likely penalize assisting suicide under
the common law of crimes,[3] whereas Hawaii and Indiana make
causing suicide an offense but do not prohibit assisting suicide.[4]
All told, only nine states have no prohibitions regarding suicide.[5]

The accepted legal definition of suicide includes: "[s]elf-destruction; the deliberate termination of one's own life"[6]; "the act of self-destruction by a person sound in mind and capable of measuring his moral responsibility."[7] Given such a wide extension, almost any act of deliberately hastening death is apt to constitute suicide in the patient who consents, and assisting suicide or even homicide in the physician who administers.[8]

Because of mounting societal interest in the issue of elective death, a handful of legal scholars have addressed the question of a constitutional right to suicide that, under certain circumstances, would protect the right of the mentally competent adult to terminate her/his own existence.[9] The existing commentary, however, is not dispositive of the issue. The writers sharply disagree on respective interpretations of constitutional doctrine as well as the general desirability of such a right. More important, the sparse commentary that does address this precise inquiry was written long before the US Supreme Court handed down its only "right-to-die" decision in *Cruzan* v. *Director, Missouri Department of Health.*[10]

A convincing case can be made for the judicial recognition of a constitutional right to elective death and the voluntary active euthanasia that such a right would engender. As such, this paper will present a straightforward legal and ethical examination of the issue in light of applicable constitutional theory. It will begin with a brief overview of the standards applied by the Supreme Court in determining the extent to which state or federal governments may interfere with individual liberty. The paper will then examine the origins and growth of constitutional privacy, and show how the erstwhile apparent applicability of privacy to the question of self-directed death was upset by the Courts' recent opinion in *Cruzan*. Nevertheless, a further examination of constitutionally protected "fundamental human rights" reveals that elective death should be treated on par with traditionally recognized fundamental rights. The chapter concludes with a brief consideration of the scope and application of the right to suicide.

Standards of Judicial Review

The Supreme Court has traditionally adopted two basic tests to determine whether the government may lawfully interfere with an individual's conduct. The first, which is often applied to questions of economic liberty, is whether the state has a "rational basis" for its restraint upon individual liberty.[11] Only if the state's regulation is so unrelated to legitimate state interests as to be arbitrary will the individual's sphere of conduct be safeguarded by the guarantees of personal liberty embodied in the Fifth and Fourteenth Amendments.[12] Decisions applying the "rational basis" test demonstrate that the Court typically accepts at face value the state's conclusion that its proscription rests on a legitimate government interest.[13]

The "rational basis" test and its inherent presumption of validity will give way to a higher level of judicial review whenever a challenger to state action can demonstrate that the questioned conduct involves the exercise of a fundamental human right. Under "strict scrutiny" analysis, the state must show that its regulation or practice advances some "compelling" state interest[14]—an "overriding" end whose value is sufficiently great as to justify the limitation of fundamental constitutional values.[15] Moreover, the state must show that its challenged infringement constitutes the least restrictive means available to sustain its compelling purpose.[16]

A number of recent decisions, however, suggest that the Supreme Court has somewhat acquiescently adopted yet a third, intermediate tier of judicial scrutiny.[17] In such cases, the Court has given very little deference to the legislature when reviewing regulatory classifications, but has employed neither the traditional "rational basis" test nor the "compelling interest" standard.[18] The "middle tier" approach has appeared most obviously in the judicial review of legislative classifications based on gender and illegitimacy,[19] and the Court has upheld such classifications only when "substantially related to an important governmental objective."[20]

Constitutional Privacy

The wall of privacy "was thrown up in great haste, from a miscellany of legal rock and stone, on two occasions and so in two parts: the first was the great Warren and Brandeis article of 1890 on the right to privacy at common law; the second was the *Griswold* decision in 1965 on the same (or a corresponding) right in constitutional law."[21] It does not appear that there has ever been a time in Anglo-American jurisprudence that a right to privacy has not been given at least implicit recognition. Among the most ancient forms of action at the common law were those that recognized a form of relief against nonconsensual invasions of bodily integrity and peace of mind.[22] The significance of the Warren and Brandeis article and the *Griswold* decision lies in the fact that both attempt to articulate explicitly a guarantee of privacy that had formerly been understood only implicitly. The concept of privacy "has had an expansive, some would say invasive, history in American law and, more than law, in the American consciousness, in American culture, and in American institutions."[23] Yet, the attempt to formulate and specifically define the parameters of the right to privacy is difficult.

Harlan's dissent in *Poe* v. *Ullman* forcefully articulated the thesis that

> the full scope of the liberty guaranteed by the Due Process Clause cannot be found in or limited by the precise terms of the specific guarantees elsewhere provided in the Constitution. This 'liberty' is not a series of isolated points pricked out in terms of the taking of property; the freedom of speech, press, and religion; the right to keep and bear arms; the freedom from unreasonable searches and seizures; and so on. It is a rational continuum which, broadly speaking, includes a freedom from all substantial arbitrary impositions and purposeless restraints [citations omitted] and which also recognizes, what a reasonable and sensitive judgment must, that certain interests require particularly careful scrutiny of the state needs asserted to justify their abridgment.[24]

Tribe has observed that "[t]he history of the framing and ratification of the Constitution and of the Bill of Rights leaves little doubt about the correctness of Justice Harlan's proposition."[25] Indeed, the Ninth Amendment, which provides that "[t]he enumeration in the Constitution, of certain rights, shall not be construed to deny or disparage others retained by the people"[26] was specifically introduced by James Madison in response to the argument that the enactment of a Bill of Rights might dangerously suggest "that those rights which were not singled out, were intended to be assigned into the hands of the General Government, and were consequently insecure."[27] Thus, the Ninth Amendment provides a fertile source of law that recognizes the existence of fundamental but unmentioned rights. As Justice Story observed: the "Bill of Rights presumes the existence of a substantial body of rights not specifically enumerated but easily perceived in the broad concept of liberty and so numerous and so obvious as to preclude listing them."[28]

Griswold v. *Connecticut*[29] sought to lend some degree of clarity to the otherwise cloudy specter of a constitutional right to privacy that protects the right of the individual to be secure from governmental intrusion into matters of a profoundly personal and intimate nature. In striking down as unconstitutional a state criminal statute that prohibited the use of contraceptives, the Supreme Court held that the government has no business intruding on the intimacies of the marital relationship: "We deal with a right of privacy older than the Bill of Rights . . . Marriage is a coming together for better or for worse, hopefully enduring and intimate to the degree of being sacred."[30] Writing for the majority, Justice Douglas observed a number of previous decisions that had recognized peripheral constitutional rights that were entailed but not expressly mandated by the explicitly guaranteed rights.[31] This analysis led Douglas to the conclusion that "specific guarantees in the Bill of Rights have penumbras, formed by emanations from those guarantees that help give them life and substance."[32] As such, certain explicitly stated constitutional guarantees—most

notably, the First, Third, Fourth, Fifth, and Ninth Amendments—
give rise to unstated "zones of privacy" that the government may
not invade. Among these zones of privacy is the intimacy endemic
to the marital relationship.

Goldberg's concurrence in *Griswold* reiterated the view that
"the concept of liberty protects those personal rights that are
fundamental, and is not confined to the specific terms of the Bill
of Rights"[33] and that "the Framers did not intend that the first
eight amendments be construed to exhaust the basic and fundamental rights which the Constitution guaranteed to the people."[34]
Rather than locating the right to privacy as an emanation from a
specific set of explicitly stated guarantees as Douglas had done,
Goldberg concluded that "the right of privacy is a fundamental
personal right, emanating from the totality of the constitutional
scheme under which we live."[35] Of even greater significance,
Goldberg cited Brandeis' famous dissent in *Olmstead* v. *United
States* as a comprehensive summary of "the principles underlying
the Constitution's guarantees of privacy"[36]:

> The protection guaranteed by the (Fourth and Fifth) Amend-
> ments is much broader in scope. The makers of our
> Constitution undertook to secure conditions favorable to
> the pursuit of happiness. They recognized the significance
> of man's spiritual nature, of his feelings and of his intellect.
> They knew that only a part of the pain, pleasure and
> satisfactions of life are to be found in material things. They
> sought to protect Americans in their beliefs, their thoughts,
> their emotions and their sensations. They conferred, as
> against the Government, the right to be let alone—the most
> comprehensive of rights and the right most valued by civi-
> lized men.[37]

Goldberg's reverence for the Brandeisian formulation of
privacy as "the right to be let alone" is significant. *Griswold* was
the first Supreme Court decision to explicitly announce the existence of the constitutional right to privacy. Yet, as Goldberg acknowl-

edges, this central notion of privacy as "the most comprehensive of rights" echoes a theme that reverberates throughout all of Anglo-American jurisprudence. The purpose of the Bill of Rights is to ensure that the individual has certain rights against the government. The theme is distinctly antimajoritarian: to protect the right of the individual to pursue her/his own vision of the good life, free from unwanted governmental intrusion, insofar as the exercise of that liberty does not significantly encroach on the welfare of others. It is this specific right of the individual to control her/his own destiny that lies at the foundation of democracy. The Bill of Rights exists as a representative set of explicitly stated guarantees that arise as particular manifestations of the more rudimentary concept of self-sovereignty.

The scope of constitutional privacy was expanded and refined by a lengthy series of decisions following *Griswold.* In the historic *Roe* v. *Wade,* for example, Douglas argued in concurrence that three overlapping areas of privacy are subsumed under the more rudimentary albeit nebulous concept of liberty:

1. "[T]he autonomous control over the development and expression of one's intellect, interests, tastes, and personality"[38];
2. "[F]reedom of choice in the basic decisions of one's life respecting marriage, divorce, procreation, contraception, and the education and upbringing of children"[39]; and
3. "[T]he freedom to care for one's health and person, freedom from bodily restraint or compulsion, freedom to walk stroll, or loaf."[40]

By the time of the 1977 decision of *Carey* v. *Population Services International,* it was clear that even though the Supreme Court had not thoroughly delineated the outer reaches of privacy, the right to personal privacy would be invoked to protect "the interest in independence in making certain kinds of important decisions"[41]—specifically, those decisions that are intimately personal and essential to the control of one's own destiny.

Many lower courts have utilized privacy as a means of protecting the right of patients to refuse or suspend artificial life support. Thus, in *Superintendent of Belchertown State School* v. *Saikewicz*,[42] the Massachusetts Supreme Court upheld the right of a guardian ad litem to refuse medical treatment on behalf of a severely retarded sixty-seven year old leukemia patient. The court acknowledged the state's "implicit recognition" that "a person has a strong interest in being free from nonconsensual invasions of his bodily integrity"[43] and held:

> of even broader import, but arising from the same regard for human dignity, and self-determination, is the unwritten constitutional right of privacy. . . . As this constitutional guaranty reaches out to protect the freedom of a woman to terminate pregnancy under certain conditions [citing *Roe* v. *Wade*], so it encompasses the right of a patient to preserve his or her right to privacy against unwanted infringements of bodily integrity in appropriate circumstances.[44]

Similarly, in the far-reaching decision of *Bouvia* v. *Superior Court (Glenchur)*,[45] the court employed privacy to permit a mentally competent, nonterminally ill quadriplegic to compel the removal of a nasogastric feeding tube. The trial court had previously denied the petitioner's request on the ground that her ostensible desire to commit suicide was not a bona fide exercise of her right to privacy. Yet the appellate court reversed, holding that "[i]f a right exists, it matters not what 'motivates' its exercise"[46] and that "a desire to terminate one's life is probably the ultimate exercise of one's right to privacy."[47]

Judicial De-Activism: Tightening the Reins

Given the law sketched briefly above, it would initially appear as though the right to privacy were itself sufficiently broad as to encompass the right of a mentally competent adult to terminate her/his own existence. But lower court precedent aside, the Supreme Court has made it clear that privacy does not protect all important personal decisions. Thus, in *Village of Belle Terre* v. *Boraas,*[48] a municipal zoning ordinance that infringed on the choice of one's living companions was allowed to stand against the charge that it violated a fundamental constitutional right. More important, in *Bowers* v. *Hardwick,*[49] a narrow majority held that there is no constitutional right to "homosexual sodomy." Moreover, in *Cruzan* v. *Director, Missouri Department of Health*[50]—the Supreme Court's only "right-to-die" decision—the Court specifically declined to invoke constitutional privacy.

In *Cruzan*, the parents of a patient in a persistent vegetative state sought to compel the termination of artificial nutrition and hydration on behalf of the patient. The Supreme Court of Missouri had reversed a lower court decision granting the relief requested, and held instead that (1) the guardians did not have authority to compel the withdrawal of hydration and nutrition, and (2) evidence of the patient's wishes was inherently unreliable and insufficient to support the guardian's claim for substituted judgment. On appeal, the question at the Supreme Court level was "simply and starkly whether the United States Constitution prohibits Missouri from choosing the rule of decision which it did."[51]

In cautiously crafting an unusually narrow holding, the Court forewarned that it would not attempt to cover every possible phase of the subject. Yet, "for purposes of this case, we assume that the United States Constitution would grant a competent person a constitutionally protected right to refuse lifesaving hydration and nutrition."[52] Nevertheless, the Court also found that the Constitution did not forbid Missouri from requiring that evidence of the incompetent's wishes as to the withdrawal of treatment be proved by clear and convincing evidence. As such, the Missouri decision was affirmed.

Justice O'Connor wrote separately to emphasize the limited reach of the decision. The *real* battle in *Cruzan* was solely ideological and was pitched between Scalia's brusque concurrence and a vigorous four-person dissent on the question of "fundamental rights" analysis.

As the Court in *Cruzan* carefully observed, liberties designated as "fundamental human rights" receive a heightened level of judicial protection. State or federal legislation infringing such rights will not pass constitutional muster unless it can be shown that the legislation in question is necessary to advance some compelling state interest and is narrowly drawn so as to constitute the least restrictive means available to sustain that compelling purpose.[54] In seeking to ascertain which rights might legitimately be deemed "fundamental," the Supreme Court has typically employed the rather nebulous rubric of two landmark decisions. In *Palko* v. *Connecticut,* it was said that this category includes those fundamental liberties that are "implicit in the concept of ordered liberty," such that "neither liberty nor justice would exist if [they] were sacrificed."[55] However, a different description of fundamental liberties can be found in *Moore* v. *City of East Cleveland,* where they are characterized as those liberties that are "deeply rooted in this Nation's history and tradition."[56]

Entrenched behind such inhospitable terrain, Justice Scalia asserted that there simply is no constitutionally recognized right to any form of elective death. His initial concern was that the federal courts "have no business in this field"[57] as the Due Process Clause protects only those deprivations of liberty traditionally protected against state interference. Since "suicide" *per se* has long been anathmatic to the law, there can be no constitutionally protected fundamental human right to die, as "there is no significant support for the claim that a right to suicide is so rooted in our tradition that it may be deemed 'fundamental' or 'implicit in the concept of ordered liberty.'"[58] But Scalia also refused to allow any differentiation between species of self-directed death: "American law has always accorded the State the power to prevent, by force if necessary, suicide—including suicide by refusing to take appropriate measures to preserve one's life."[59] In this sense, "[s]tarving oneself to death is no different from putting a gun to one's temple as far as the common law definition of suicide is concerned; the cause of death in both cases is the suicide's conscious decision to 'pu[t] an end to his own existence.'"[60]

The dissent, however, aptly countered that Due Process concerns regarding the protection of individual liberties were clearly involved in the case. Whereas Scalia had strategically fixed quite rigidly on the ancient proscription of "suicide" in determining whether there is a fundamental human right to choose to die, the dissent offered a much broader perspective: "The right to be free from medical attention without consent, to determine what shall be done with one's own body, *is* deeply rooted in this Nation's traditions... This right has long been 'firmly entrenched in American tort law' and is securely grounded in the earliest common law."[61] Properly:

> It is against this background of decisional law, and the constitutional tradition which it illuminates, that the right to be free from unwanted life-sustaining medical treatment must be understood. That right presupposes no abandonment of the desire for life. Nor is it reducible to a protection against batteries undertaken in the name of treatment, or to a guarantee against the infliction of bodily discomfort. Choices about death touch the core of liberty. Our duty, and the concomitant freedom, to come to the terms with the conditions of our own mortality are undoubtedly "so rooted in the traditions and conscience of our people as to be ranked as fundamental" [citations omitted] and indeed are essential incidents of the unalienable rights to life and liberty endowed to us by our Creator.[62]

It is important to note that the majority in Cruzan "assumed" that a competent person has a "constitutionally protected liberty interest"[63] in declining unwanted medical treatment. Hidden away in a footnote, the Court cryptically suggested that "[a]lthough many state courts have held that a right to refuse treatment is encompassed by a generalized right of privacy, we have never so held. We believe this issue is more properly analyzed in terms of a Fourteenth Amendment liberty interest."[64] But what are the parameters of this "liberty interest," and more important, why was the constitutional right to privacy (the erstwhile obvious choice of doctrine) not brought to bear?

As the Court itself has noted, the exact boundaries of the right to privacy are somewhat ill-defined. At least one commentator has derisively labeled the right "unprincipled" because it cannot be derived by reference to neutral constitutional precepts.[65] Under this view, the choice of fundamental values initially listed in *Griswold* was unjustified because in the absence of specific constitutional preference, there is simply no principled way to prefer one value over another. Thus, it is left to the Justices, without guidance from the text of the Constitution, to determine which liberties may be infringed on and which may not. It has been further suggested that the rights traditionally protected by the right of privacy are "family-oriented" and that the Court did not want to extend the mantle of privacy beyond this realm. Following this tack: "[t]he right to privacy does not include a right to refuse life-sustaining medical treatment because the decision to refuse medical treatment is not an element of marriage, procreation, and family."[66] Still, a more palpable explanation of the Court's reluctance in *Cruzan* to extend the reach of privacy lies in the claim that the conservative Court merely wanted to tighten the reins of privacy. Thus, "in recent years the Court's conservative majority has shown a growing restiveness with the expansive privacy rulings of the Warren and Burger years, and it may be that Cruzan offer[ed] a chance to trim the privacy right without having to use abortion—always a volatile issue—to do so."[67]

Both the "liberty interest" at issue in *Cruzan* and the right to privacy are rooted in the Due Process Clauses of the Fifth and Fourteenth Amendments. A literal reading of the clauses focuses on "the processes by which life, liberty, or property is taken."[68] Under *procedural* due process review, no right classified as a liberty interest can be infringed by the state or federal governments unless certain processes or preconditions are met. The relevant state interests involved determine the amount or type of process actually due in protecting the liberty interest. Procedural review is limited in scope and guarantees only that there is a fair decision making process before the government actually impairs a person's life, liberty, or property. The right to privacy, however, developed from a line of decisions that interpreted the Due Process Clauses as having substantive content. Under *substantive* review, the Court is concerned with the constitutionality of the underlying rule rather than the fairness of the process by which the rule is applied. Accordingly, certain avenues of substantive due process review have met with intense criticism as the judicial determination of the legitimacy of the "substance" of a law threatens to usurp the proper function of the legislature. Under this line of argument, it would appear that the Court in *Cruzan* was actually reluctant to expand the sphere of privacy precisely because the doctrine itself—especially under the quasi-legislative "framework" of *Roe* v. *Wade*—is tainted with the stigma of being judge-made law. Yet if the right to die can be discovered via some other, less controversial, avenue, then the Court can "recognize" the existence of the right with a clear conscience.

Fundamental Rights

Little can be said toward describing the nature of a fundamental right, and it might be argued that fundamental rights analysis is nothing more than the contemporary recognition that a higher law limits the restrictions of liberty that a temporal government might impose. Historically, it is clear that prior to the Civil War, the US Supreme Court had held that the provisions of the Bill of Rights served to check the powers of the federal government, but were not applicable to state and local governments. However, after the passage of the Fourteenth Amendment in 1868, the Court began a process of incorporating the various guarantees afforded by the Bill of Rights into that Amendment in order to limit the

powers of states and municipalities. Today, nearly all of the Bill of
Rights has been incorporated into the Fourteenth Amendment and is
made applicable to the states in terms of fundamental rights that may
not be invaded. This list includes—but is not necessarily limited to—
all of the First Amendment provisions regarding the freedom of
religion,[69] speech,[70] press,[71] assembly,[72] and petition;[73] the Fourth
Amendment's regulation of searches and seizures;[74] the Fifth
Amendment's double jeopardy[75] and self-incrimination provisions;[76]
the Sixth Amendment rights of criminal process;[77] and the Eighth
Amendment's admonition against cruel and unusual punishment.[78]

In addition to the rights incorporated from the express provisions of the Bill of Rights, the Supreme Court has recognized
a second and more limited sphere of fundamental rights that are
implied but not expressly articulated in the Constitution's text.
This enumeration includes: freedom of association,[79] the right to
equal participation in the political process,[80] the right to interstate
travel,[81] the right to basic fairness in the criminal process[82] and in
procedures regarding claims against governmental deprivations
of life, liberty, or property,[83] and the right to privacy.[84]

The Right to Suicide

An analysis of the constitutional right to privacy reveals that
privacy protects those decisions—such as marriage, procreation,
contraception, abortion, family relationships, and child-rearing and
education—that are personal, intimate, and relate to the control of
one's own destiny. Certain "zones of privacy"[85] exist under the
Constitution that embody a promise that a "private sphere of individual liberty will be kept largely beyond the reach of government."[86]
Within that sector of individual liberties are those decisions that form
a central part of an individual's life and concern "unusually important decisions that will affect his own, or his family's destiny."[87]
"[T]he concept of privacy embodies the 'moral fact that a person
belongs to himself and not others nor to society as a whole.'"[88]

Although the outer reaches of privacy have not been thoroughly delineated by the Supreme Court, the Justices have treated the various particular "rights" of privacy as corollary liberties flowing from the deeper principle that the individual should have unfettered liberty over her/his own life and person insofar as the exercise of such liberty does not seriously encroach on the welfare of others. Accordingly, the judicially recognized "rights" of privacy represent a cluster of rights contained within the Constitution's broader protection of individual liberty. The Court has also recognized a constitutionally protected liberty interest in declining unwanted medical treatment and it has traditionally offered zealous protection over interests that are of paramount concern to the individual—freedom of expression, association, religious belief, and travel; security from unreasonable searches and seizures; protection from cruel and unusual punishment; rights of basic fairness in criminal and related proceedings; and even the right to equal participation in the political process. The Constitution protects this panoply of fundamental human rights precisely because such rights concern matters that are profoundly personal and relate so integrally to the basic freedom of the individual to shape and define her/his own destiny. That decision that fits this criterion more closely than any other is the decision whether or not to continue living.

Alternatively, those rights deemed "fundamental" by the Court and entitled to heightened judicial protection stem from a core commitment to an inchoate right of personal autonomy that preserves the right of the individual to master her/his own fate, free from unwarranted government intrusion into personal choices that will harm no one else. Moreover, this right satisfies the traditional tests applied by the Court in fundamental rights analysis. The right of basic self-sovereignty is conceptually "implicit in the concept of ordered liberty" and is undeniably "deeply rooted in this Nation's history and tradition." Indeed, the decision to exit life by one's own decree is more "fundamental" to the concepts of autonomy, freedom, and liberty than any other, for pivotal to the control of one's life is the choice of electing to forgo continued life. Hence, constitutional commitments to the protection of individual liberty protect the right of the individual—at least under certain circumstances—to terminate her/his own existence.

As the plurality in *Cruzan* aptly shows, the lines of battle can be narrowed to one decisive point: Anglo-American law is anchored on the premise of thorough-going self-determination, yet suicide has always been held in strictest anathema. In the proverbial "no man's land" between fronts lie numbers of seriously ill, mentally competent adults who have rationally concluded that life is no longer worth living and yet lack the wherewithal to end their own suffering. As the political and ideological war rages, more hapless patients and their families fall victim. A satisfactory accord must be struck before the situation can be ameliorated.

If we venture beyond the immediate surface of the controversy, it becomes apparent that the exaltation of self-sovereignty reflects the belief that individual freedom is one of the principle goods in life. Society's interest in the prevention of suicide represents, out of lesser concerns, the conviction that human life is of such penultimate value that it must never be cast away without sufficient warrant. Yet an examination of the common law origins of the admonition against self-directed death reveals that the edict rests almost exclusively on dubious theological dogma that will neither compel the assent of those who do not accept the traditional eschatology nor successfully scale the wall of separation between church and state.[89] Moreover, legitimate contemporary rationales for the proscription of suicide presuppose that the life being preserved is of present or future value to the individual. A conception of life *per se* devoid of any individual and personal interest in living, is an empty abstraction. The state should take every reasonable precaution to avoid the unnecessary and regrettable loss of life, but it should never insist that a competent adult remain alive against her/his own decided interests.

If the Supreme Court of the United States were to recognize suicide as a fundamental human right under the Constitution, then suicide would be entitled to "strict scrutiny" protection. This would mean that state and federal governments could not infringe this right unless the legislation in question were necessary to advance some compelling state interest and were narrowly drawn so as to constitute the least restrictive means available of sustaining that purpose. The preservation of life is legitimately construed as a compelling interest of the government to the extent that the individuals whose lives are at issue have some considered personal interest in living; reciprocally, the state does not have a legitimate interest in seeking to compel the continued existence of all persons at all costs. Flat proscriptions of suicide that prohibit all deliberate acts of self-destruction are not narrowly tailored so as to provide the least restrictive means of promoting the state's compelling purpose. As such, judicial recognition of the constitutional right to suicide would signal the decline of overly-broad statutes that sweepingly condemn all deliberate acts of self-destruction or the aiding and abetting of such acts. Nevertheless, recognition of such a right would not forbid legislation that prescribes the manner in which the right is exercised. In this way, incompetents and minors may still be restrained from senseless acts of self-slaughter. Moreover, the traditional parade of horrors so frequently conjured by the assailants of choice in dying is sorely misplaced. Individuals who want to die and are capable of carrying out their wishes are not likely to be either deterred or facilitated by a change in the legal status of suicide. In this sense, flat proscriptions of self-willed death—and assisting in such acts—effectively hamper that one class of persons who should be allowed to die: terminal and seriously ill patients. Thus, when all is told, recognition of the constitutional right to suicide must be seen as the ultimate forum for the preservation of the quality and value of human life as well as respect for the sanctity of personal choice.

Application

The common law has long recognized the right of the individual to be free from nonconsensual invasions of bodily integrity.[90] Since this right has been extended to include the freedom to refuse even necessary life-saving medical treatment, the right to refuse intervention carries an implicit (and supposedly passive) right to die. Thus, patients are ordinarily permitted to refuse the application of life support apparatus or even artificial nutrition and hydration. But what about seriously ill patients who are not imminently threatened with the application of mechanical life support? Must they be forced to linger and wait for the inevitable, even if all attempts at pain management have failed? Because one cannot deliberately choose to kill oneself, the present "right-to-die" exists merely as a negative claim that permits the patient to repel invasions of bodily integrity. As a negative claim, it does not grant the patient a positive right to make claims on others, or to demand the application of procedures that might, however rationally and humanely, directly contribute to her/his demise.

The proposed constitutional right to suicide would allow the competent adult greater control over the circumstances of dying. While the law currently permits a type of passive euthanasia whereby patients can allow themselves to be overtaken by the ravages of injury or disease, the purported distinction between the illegal self-infliction of deadly harm and a permissible determination against medical intervention is tenuous, and ineluctably works to the detriment of seriously ill patients who reasonably conclude to summon death directly. The right engendered is specifically construed as a right to suicide because it would allow the patient to initiate a course of action that would directly result in death. The death-producing agency might take the form of direct and active disemployment of life-support, a lethal injection or intentional drug overdose, or even the utilization of a Kevorkian—prototype "suicide machine." As a positive right, it would extend beyond the mere negative claim involved in repelling invasions of bodily integrity. Gravely ill, competent patients who wanted to die, even if not immediately ensnared by invasive life-support apparatus, could compel the administration of lethal agency and, arguably, present certain positive claims against others. Thus, if the individual were unable to personally accomplish the decision to die, or if the process of dying could be made more humane with the intervention of others, then the patient would have the right to solicit proper assistance. The assistance of suicide, at least under these circumstances, could no longer be held criminal and tantamount to homicide, but would more aptly be regarded akin to suicide by proxy, or suicide requiring instrumental agency.

Judicial recognition of the constitutional right to suicide will not lead to abuse. A proper appreciation of the nature of a constitutionally protected fundamental right circumvents any fear of adverse societal impact. The First Amendment, for example, guarantees freedom of expression as a fundamental human right and yet one may not exercise this right so as to create a public disturbance,[91] promote obscenity,[92] or commit libel.[93] Similarly, the fundamental right to religious belief does not entitle one to exercise this prerogative in any fashion one deems fit.[94] If the proposed right to suicide were recognized, state and federal governments could institute sufficient regulatory safeguards to prevent even the risk of impropriety. The state's only requirement would be to show that the regulation in question was necessary to further the compelling interest in preventing the unwarranted loss of life, and was narrowly drawn as to be minimally restrictive. Hence, following the Supreme Court's abortion rulings as an example, the state could demand that the party assisting a suicide be licensed,[95] and that the faculty and proxy involved keep proper medical records.[96] The state could require the written consent of the patient,[97] or even the presence of a second licensed proxy.[98] Moreover, given the unique nature of suicide, the state might also impose a certain minimal waiting period, complete disclosure of all of the patient's options, and more important, even a preliminary psychiatric evaluation. The law surrounding the fundamental right to marry evinces the fact that the state need not permit the suicide of minors or incompetents.[99]

The right to elective death would be treated as a positive right in much the same way that abortion may be considered a positive right. The right to suicide would generate a freedom from undue government interference, and a corresponding right to solicit proper assistance. Of course, a woman's right to elect an abortion does not create an obligation on the part of any physician to perform it. Yet provided that she locates an appropriate and willing practitioner, she is at liberty to solicit such services free of unnecessary restraint. In this way, a request for assisted suicide would be treated on par with any of the more traditional medical procedures. Accordingly, one could not arbitrarily compel one's attending physician to facilitate one's suicide, were such an act in contravention of the physician's considered assessment of the patient's prognosis, beyond the physician's declared area of practice, or even simply beyond the physician's sheer willingness to participate. However, were it the case that one's treating physician concurred in one's judgment that the option of death presented the most rational and advantageous course of action in light of the circumstances, one could solicit necessary assistance. The physician, medical assistant, or other specified proxy would, in turn, be immune from criminal or civil prosecution provided that the patient granted informed consent and the proxy exercised that degree of care in the undertaking necessary to satisfy prevailing professional standards.

Consonant with this position, Dr. Jack Kevorkian has convincingly argued that medicine should recognize a new specialty designed to derive human benefit from death and dying.[100] "Obitiatrists" could accordingly lend a comforting hand to those slipping into death, and extract knowledge from the dying to help those left behind.

Kevorkian's argument is well-founded. Given the present state of the law and medicine, individuals with progressive, degenerative diseases must elect either to take their own life during an early stage of the disease when they are still physically and mentally capable—often prematurely and without optimal skill—or, suffer the grim realities of a lingering and dehumanizing decline. If physician-assisted death were an accepted practice, patients could enjoy their last days on earth, secure in the knowledge that they will not be forced, against their will, to endure hopeless pain and degeneration. Obitiatrists could provide counseling for the patient and family members to allow all persons involved to come to terms with death. Such counseling would enable the obitiatrist to isolate and refuse services to persons whose request for assisted suicide was prompted by illegitimate ends or psychological maladies alone. Federal and state regulatory agencies could monitor professionally- licensed practitioners. Physicians could ensure that patients would meet death with dignity and self-mastery, and the implementation of "suicide machines" would minimize the need for direct intervention. Planned death would also offer the possible benefits of organ and tissue donation as well as advances in medical research.

Concluding Remarks

The simple argument from mercy presents a powerful moral case for the legalization of physician-assisted death. The foregoing pages, despite their brevity, evince that there is also a convincing argument in support of judicial recognition of a constitutional right to suicide. If one impartially examines the available literature, it becomes clear that the arguments opposing the legalization of suicide and physician-aided death almost invariably hinge on logically fallacious "slippery-slope" objections or dogmatic pontifications regarding the unqualified immorality of any form of elective death. Yet at bottom, much of the debate can be distilled into a single item of contention. Although the proponents of voluntary euthanasia and assisted suicide would not force an individual to die against his or her will, neither would they compel a mentally competent adult to remain alive against his or her wishes and best interests. Opponents of legalized voluntary euthanasia and suicide, conversely, deny the right of the individual to choose to die. However, if it is wrong to kill a person against his or her will, it would seem equally wrong—under a wide variety of circumstances—to force the competent adult to remain alive. Life is simply not worth living at all costs, and respect for human dignity demands respect for the choice to die.

Notes and References

[1]Marzen, O'Dowd, Crone, and Balch, Suicide: A Constitutional Right? *24 Duq. L. Rev.* 1, 97 (1985).

[2]*Ibid.*

[3]*Ibid.* But *see*: Wayne R. LaFave and Austin W. Scott, Jr., *Handbook on Criminal Law,* West Publishing Co., St. Paul, MN, 2nd ed. (1986) p. 649.

[4]Marzen, O'Dowd, Crone and Balch, p. 98.

[5]*Ibid.*

[6]Black's Law Dictionary 1434, 6th ed. (1990).

[7]*Webster's New Twentieth Century Dictionary 1822,* 2nd ed., 1979.

[8]Glanville Williams, *The Sanctity of Life and the Criminal Law,* Alfred A. Knopf, New York (1972), p. 318.

[9]Alan Sullivan, A Constitutional Right to Suicide, in M. Pabst Battin and David J. Mayo, eds., *Suicide: The Philosophical Issues,* St. Martin's Press, New York (1980), pp. 229-253; David A. J. Richards, Constitutional Privacy, the Right to Die and the Meaning of Life: A Moral Analysis, *22 Wm. and Mary L. Rev.* 327 (1981); H. Tristram Engelhardt and Michele Malloy, Suicide and Assisting Suicide: A Critique of Legal Sanctions, *30 Sw. L. J.* 1003 (1982); Marzen, O'Dowd, Crone, and Balch, Suicide: A Constitutional Right? *24 Duq. L Rev.* 1 (1985).

[10]*Cruzan* v. *Director, Missouri Department of Health,* 110 S. Ct. 2841 (1990).

[11]Alan Sullivan, A Constitutional Right to Suicide, in M. Pabst Battin and David J. Mayo, eds., *Suicide: The Philosophical Issues,* St. Martin's Press, New York (1980), pp. 229–253, 232, citing *Williamson* v. *Lee Optical. Inc.,* 348 U. S. 483, 488 (1955).

[12]Sullivan, p. 232, citing *Kelley* v. *Johnson,* 425 U. S. 238 (1976).

[13]Sullivan, p. 232.

[14]*Ibid.,* citing *Roe* v. *Wade,* 410 U. S. 113, 153 (1973); *Kramer* v. *Union Free School District,*395 U. S. 621, 627 (1969); *Shapiro* v. *Thompson,* 394 U. S. 618, 634 (1969).

[15]John E. Nowak and Ronald D. Rotunda, *Constutional Law,* West Publishing Co., St. Paul, Minn., 4th ed. (1991), p. 575.

[16]Sullivan, p. 232, citing *Shelton* v. *Tucker,* 364 U. S. 479, 488 (1960).

[17]Nowak and Rotunda, p. 576.

[18]*Ibid.,* citing *Alabama State Federation of Teachers* v. *James,* 656 F. 2d 193, 195 (5th Cir. 1981); *Torres* v. *Portillos,* 638 P. 2d 274, 276 (Colo. 1981); *Lujan* v. *Colorado State Bd. of Educ.,* 649 P. 2d 1005,1015 n. 9 (Colo. 1982).

[19]Nowak and Rotunda, p. 576.

[20]*Clark* v. *Jeter,* 486 U. S. 456 (1988); *Craig* v. *Boren,* 429 U. S. 190, 197 (1976).

[21]Thomas Gerety, Redefining Privacy, 12 Harv. C.R.—C.L.L. Rev. 233 (1977) citing Warren and Brandeis, The Right to Privacy, 4 *Harv. L. Rev.* 193 (1890) and *Griswold* v. *Connecticut,* 381 U. S. 479 (1965).

[22]I de S et ux. v. W de S, 1348, Y.B. Lib. Assis. f. 99, pl. 60; 1366, Y.B. 40 Edw. III 40, pl. 19; *Smith* v. *Newsam,* 1674. 3 Keb. 283, 34 Eng. Rep. 722.

[23]Gerety, p. 234, citing to Kalven, Privacy in Tort Law—Were Warren and Brandeis Wrong? 31 *Law and Contemp. Prob.* 326 (1966).

[24]*Poe* v. *Ullman,* 367 U. S. 497, 543 (1961) (Harlan, J. dissenting).

[25]Laurence H. Tribe, *American Constitutional Law,* Foundation Press, Mineola, NY (1978), p. 570.

[26]US Const. amend IX.

[27]1 Annals of Cong. 439, Gales and Seaton ed. (1834).

[28]3 Story, *Commentaries on the Constitution of the United States,* 715–716 (1833); *See also* 2 Story, *Commentaries on the Constitution of the United States,* 626–627, 651, 5th ed. (1891).

[29]*Griswold* v. *Connecticut,* 381 U. S. 479 (1965).

[30]381 U. S. at 486.

[31]Thus, the right to educate one's children as one sees fit was made applicable to the States by the force of the First and Fourteenth Amendments under *Pierce* v. *Society of Sisters,* 268 U. S. 510

(1925). Under *Meyer* v. *Nebraska,* 262 U. S. 390 (1923), the same dignity was given the right to study the German language in a private school. The right of freedom of speech and press was found to include not only the right to utter or to print, but also the right to distribute, the right to receive, and the right to read by *Martin* v. *Struthers,* 319 U. S. 141(1943). Freedom of inquiry, freedom of thought, and freedom to teach was brought under the auspices of the First Amendment by *Wieman* v. *Updegraff,* 344 U. S. 183 (1952) and this was extended to the freedom endemic to the entire university community by *Sweezy* v. *New Hampshire,* 354 U. S. 234 (1957). Douglas added: "Without those peripheral rights the specific rights would be less secure." 381 U. S. at 482–483.

[32]381 U. S. at 484.

[33]381 U. S. at 486.

[34]381 U. S. at 490.

[35]381 U. S. at 494, citing *Poe* v. *Ullman,* 367 U. S. 497, 521 (Douglas, J., dissenting).

[36]381 U. S. at 494.

[37]*Ibid.,* citing *Olmstead* v. *United States,* 277 U. S. 438, 478 (1928) (Brandeis, J. dissenting).

[38]*Roe* v. *Wade,* 410 U. S. 113, 211 (1973).

[39]*Ibid.*

[40]410 U. S. at 213.

[41]*Carey* v. *Population Services International,* 431 U. S. 678, 684 (1977), citing *Whalen* v. *Roe,* 429 U. S. 589 (1977).

[42]*Superintendent of Belchertown State School* v. *Saikewicz,* 373 Mass. 728, 370 N.E. 2d 417 (1977).

[43]373 Mass. at 738–739, 370 N.E. 2d at 424.

[44]373 Mass. at 739, 370 N.E. 2d at 424.

[45]*Bouvia* v. *Superior Court (Glenchur),* 179 Cal. App. 3d 1127, 225 Cal. Rptr. 297 (Ct. App. 1986), rev. den. (Cal. June 5, 1986).

[46]179 Cal. App. 3d at 1145, 225 Cal. Rptr. at 306.

[47]179 Cal. App. 3d at 1144–1145, 225 Cal. Rptr. at 306.

[48]*Village of Belle Terre* v. *Boraas,* 416 U. S. 1(1974). *But see: Moore* v. *East Cleveland,* 431 U. S. 494 (1977).

[49]*Bowers* v. *Hardwick,* 478 U. S. 186 (1986).

[50]*Cruzan* v. *Director, Missouri Department of Health,* 110 S. Ct. 2841 (1990).

[51]110 S. Ct. at 2851.

[52]110 S. Ct. at 2852.

[53]*Roe* v. *Wade*, 410 U. S. 113 (1973); *Kramer* v. *Union Free School District*, 395 U. S. 621 (1969); *Shapiro* v. *Thompson*, 394 U. S. 618 (1969).

[54]*Shelton* v. *Tucker*, 364 U. S. 479 (1960).

[55]*Palko* v. *Connecticut*, 302 U. S. 319, 325–326 (1937).

[56]*Moore* v. *East Cleveland*, 431 U. S. 494, 503 (1977).

[57]*Cruzan* v. *Director, Missouri Department of Health*, 110 S. Ct. 2841, 2859 (1990) (Scalia, J., concurring).

[58]110 S. Ct. at 2860 (Scalia, J., concurring) citing Marzen, O'Dowd, Crone, and Balch, Suicide: A Constitutional Right? *24 Duq. L Rev.* 1, 100 (1985), quoting *Palko* v. *Connecticut*, 302 U. S. 319, 325 (1937).

[59]110 S. Ct. at 2859 (Scalia, J., concurring).

[60]110 S. Ct. at 2861 (Scalia, J., concurring) citing 4 W. Blackstone, *Commentaries* 189.

[61]110 S. Ct. at 2865 (Brennan, J., dissenting).

[62]110 S. Ct. at 2885 (Stevens, J., dissenting).

[63]110 S. Ct. at 2851.

[64]110 S. Ct at 2851 n. 7.

[65]Robert Bork, Neutral Principles and Some First Amendment Problems, *47 Ind. L. J.* 1 (1971).

[66]Edward A. Lyon, The Right to Die: An Exercise of Informed Consent, Not an Extension of the Constitutional Right to Privacy, *58 U. Cin. L. Rev.* 1367, 1388 (1990).

[67]Ross Nankivell, This Far and No Further: Is There a Constitutional Right to Die? *A.B.A. J.* April 1990, 66, 67.

[68]*Bowers* v. *Hardwick*, 478 U. S. 186,191(1985).

[69]*Cantwell* v. *Connecticut*, 310 U. S. 296 (1940); *Everson* v. *Board of Education*, 330 U. S. 1 (1947).

[70]*Gitlow* v. *New York*, 268 U. S. 652 (1925); *Fiske* v. *Kansas*, 274 U. S. 380 (1927); *Stromberg* v. *California*, 283 U. S. 359 (1931).

[71]*Near* v. *Minnesota*, 283 U. S. 697 (1931).

[72]*DeJonge* v. *Oregon*, 299 U. S. 353 (1937).

[73]*Ibid.*, see also: *Hague* v. *CIO*, 307 U. S. 496 (1939); *Bridges* v. *California*, 314 U. S. 252 (1941).

[74]*Wolf* v. *Colorado*, 338 U.S. 25 (1949); *Mapp* v. *Ohio*, 367 U.S.643(1961).

72   *Neeley*

[75]*Benton* v. *Maryland,* 395 U. S. 784 (1969).
[76]*Malloy* v. *Hogan,* 378 U. S. 1 (1964).
[77]*Klopfer* v. *North Carolina,* 386 U. S. 213 (1967); *In re Oliver,* 333 U. S. 257 (1948); *Gideon* v. *Wainwright,* 372 U. S. 335 (1963).
[78]*Louisiana ex. rel. Francis* v. *Resweber,* 329 U. S. 459 (1947) rehearing denied 330 U. S. 853 (1947); *Robinson* v. *California,* 370 U. S. 660 (1962) rehearing denied 371 U. S. 905 (1962).
[79]*NAACP* v. *Alabama ex. rel. Patterson,* 357 U. S. 449 (1958); *Bates* v. *Little Rock,* 361 U. S. 516 (1960).
[80]*Harper* v. *Virginia State Bd. of Elections,* 383 U. S. 663 (1966); *Carrington* v. *Rash,* 380 U.S.89 (1965).
[81]*Shapiro* v. *Thompson,* 394 U. S. 618 (1969).
[82]*Douglas* v. *California,* 372 U. S. 353 (1963); *Mayer* v. *Chicago,* 404 U. S. 189 (1971); *Bounds* v. *Smith,* 430 U. S. 817 (1977).
[83]*Youngberg* v. *Romeo,* 457 U. S. 307 (1982); *Santosky* v. *Kramer,* 455 U. S. 745 (1982).
[84]The right to privacy is generally held to include decisions affecting marriage, *Loving* v. *Virginia,* 388 U. S. 1(1967); procreation, *Skinner* v. *Oklahoma,* 316 U. S. 535 (1942); contraception, *Eisenstadt* v. *Baird,* 405 U. S. 438 (1972); abortion, *Roe* v. *Wade,* 410 U. S. 113 (1973); family relationships, *Prince* v. *Massachusetts,* 321 U. S. 158 (1944); and child rearing and education, *Pierce* v. *Society of Sisters,* 268 U. S. 510 (1925); *Meyer* v. *Nebraska,* 262 U. S. 390 (1923).
[85]*Griswold* v. *Connecticut,* 318 U. S. 479, 484 (1965).
[86]*Thornburgh* v. *American Coll. of Obst. and Gyn.,* 476 U. S. 747, 772 (1986).
[87]Fitzgerald v. Porter Memorial Hospital, 523 F. 2d 716, 719–720, (CA 7 1975), cert. den. 425 U. S. 916 (1976).
[88]*Bowers* v. *Hardwick,* 478 U. S. 186, 204 (1985) (Blackman, J., dissenting) citing *Thornburgh* v. *American Coll. of Obst. and Gyn.,* 476 U. S. 747, 777 n. 5 (1985).
[89]*See:* Norman St. John-Stevas, *Life, Death and the Law,* Indiana University Press, Bloomington, IN. (1961); Glanville Williams, *The Sanctity of Life and the Criminal Law,* Alfred A.Knopf, New York (1972).
[90]*Union Pacific Railway Co.* v. *Botsford,* 141 U. S. 250, 251(1891).
[91]*Chaplinsky* v. *New Hampshire,* 315 U. S. 568 (1942).

[92]*Miller* v. *California,* 413 U. S. 15 (1973).

[93]*Beauharnais* v. *Illinois,* 343 U. S. 250 (1952) rehearing denied 343 U. S. 998 (1952).

[94]*Jacobson* v. *Massachusettes,* 197 U. S. 11 (1905); *Zucht* v. *King* 260 U. S. 174 (1922).

[95]*See: Connecticut* v. *Menillo,* 423 U. S. 9 (1975) (per curium).

[96]*See: Planned Parenthood of Central Missouri* v. *Danforth,* 428 U. S. 52 (1976).

[97]*Ibid.*

[98]*See: Planned Parenthood Association of Kansas City, Missouri, Inc.* v. *Ashcraft,* 462 U.S.476 (1983).

[99]*See: Zablocki* v. *Redhail,* 434 U. S. 374 (1978).

[100]Jack Kevorkian, *Prescription: Medicide,* Prometheus Books, Buffalo, New York (1991), p. 203.

Physician-Assisted Suicide and Active Euthanasia

Franklin G. Miller
and John C. Fletcher

Introduction

Controversy rages about physician-assisted suicide and active euthanasia in the news media, the scholarly literature of medicine and bioethics, and the political arena of public referenda. In this chapter, after a brief description of cultural signs that favor moral experimentation in this area, we assess the relative merits of physician-assisted suicide (PAS) and physician-performed euthanasia (PPE) from an ethical perspective. The conclusion is reached that there is no good reason, all things considered, for holding that PAS is morally superior to PPE. We recommend legalizing both PAS and PPE for competent patients suffering from terminal illnesses who autonomously choose to end their lives. We argue that in order to safeguard against abuses, legalization of voluntary euthanasia must be accompanied by mandatory prior committee review. The nature of committee review is described briefly and various objections to this policy are examined critically.

A Favorable Cultural Context

Some favorable cultural signs exist suggesting readiness to learn if far more good than harm would flow from careful experiments with prevailing moral policy about when and how to die. One sign is the pace of cultural evolution toward greater choice in dying. For centuries death off the battlefield occurred at home or at accident scenes. In the United States and other developed nations, dying and the event of death have been rapidly transformed mainly into a clinical event that is controlled by decisions. In this society such decisions to permit death by forgoing medical treatment are shared between patients, physicians, relatives, or significant others. Ordinary dying in the United States is no longer a "natural" event, something that simply happens to us.

From the time of the Quinlan case in the mid-1970s to the present, "negotiated dying" or "death by decision"[1] in a clinical context has evolved into the mainstream American way of dying. Most Americans die in hospitals or nursing homes.[2] Death is prefaced typically by decisions to withhold or withdraw treatment, in a framework of "allowing to die" that eschews "killing." Choices are made by competent patients and family or other surrogates for incompetent patients, with the help and recommendations of physicians, joined by others who may know the patient clinically, such as nurses, social workers, or clergy.

In recent years Americans have had a profound moral education in the cauldron of their own experiences—in the context of families, health care, and in public controversies—about the ethical, legal, and social issues of dying. Many medical, social, and legal innovations were needed to facilitate "death by decision." Attention to the suffering and needs of dying persons in the framework of principles of hospice care has become part of American culture.[3] Numerous court decisions have upheld the rights of patients and family to forgo treatment that merely prolongs the process of dying or is not in the best interests of the patient.[4] Social and legislative changes include advance directives like the living will and durable powers of attorney for health-care[5] and protection from lawsuits for physicians who act in accord with the wishes embodied in the directives. Also, many states have enacted laws enabling legally authorized surrogates and physicians to make decisions to "allow to die" for terminally ill patients who are incapable of decisions and have no advance directives.

Some ethicists[6] see no morally relevant differences between these decisions and voluntary euthanasia. Others, such as Beauchamp and Childress,[7] hold that there are important differences but admit that there exist cases where the line is seriously blurred: e.g., forgoing treatment in a patient with a severe stroke who could live a long time, albeit with irreversible impairment and poor quality of life, aided by artificial feeding, hydration, and other measures. However, despite these ethical ambiguities, many new choices have been added for the dying and their companions who survive. Whether as a society we should move beyond allowing to die by permitting physicians to aid patients in dying, in the form of assisted suicide or active euthanasia, has become a topic of intense public debate.

Public opinion favoring physician aid in dying is a second favorable sign of the timeliness of moral experimentation. Table 1 collects public opinion research on legalizing euthanasia from 1947 to 1991[8] and shows a steady growth in approval.

With one exception, studies about euthanasia conducted among physicians cannot be interpreted with confidence. Most studies to date are marred by low response rates and problems with design. In the only well-designed study with a good response rate, a 1988 Harris poll[9] of a sample representative of US physicians found that 30% favored euthanasia if a terminally ill patient requested it (60% believed it was wrong and 4% were not sure). A 1988 survey of 7095 Colorado physicians[10] with a 31%-response rate found that 60% had attended a patient for whom they believe active euthanasia would have been justified if it were legal, and 59% would "personally have been willing to administer a lethal dose of medication." A 1991 survey[11] of 2000 randomly selected

Table 1
Public Opinion and Legalizing Euthanasia, 1947–1991

When a person has a disease that cannot be cured, do you think that doctors should be allowed by law to end the patient's life by some painless means if the patient and (his or her) family request it?

Year	Organization	Results
1947	Gallup	37% Yes, 54% No, No Opinion 9%
1950	Gallup	36% Yes, 64% No or No Opinion
1973	Gallup	53% Yes, 40% No, No Opinion 7%
1982	National Opinion Research Center	61% Yes, 34% No, Don't Know 5%
1982	Louis Harris	53% Yes, 38% No, Not Sure 8%
1986	National Opinion Research Center	66% Yes, 31% No, Don't Know 4%
1987	Louis Harris	62% Yes, 34% No, Not Sure 4%
1991	Associated Press	60% Yes, 24% No, Don't Know 8%, Not familiar 7%, Refused 1%
1991	Krc Communications	63% Yes, 28% No, Don't Know 9%
1991	National Opinion Research Center	70% Yes, 25% No, Don't Know 5%

Source: Roper Center for Public Opinion Research, University of Connecticut, 1992.

members of the Washington State Medical Society (55% response rate) found that 51% opposed and 49% supported an initiative to legalize euthanasia. This result is virtually identical to the public vote on the initiative. Given the cautions mentioned above, these studies may indicate that a moderate-sized minority of physicians are now willing to cooperate with patient requests if euthanasia is legalized.

Do some physicians in the United States secretly assist in the suicide or intervene to terminate the lives of hopelessly ill and moribund patients? The Colorado study, as well as three other anonymous polls, suggest that such acts do occur but their frequency is unknown.

A third sign of a climate favorable to moral experimentation is the appearance of articles by experienced physicians involved in the debate about euthanasia. They recognize that, despite the improvement in relief of suffering of dying patients by means of hospice care, there are and will be patients whose situations call for additional options.[13]

The Moral Coherence
of Limited, Voluntary Euthanasia

Discussion of the relative merits of PAS and PPE makes sense only if one or both of these modes of "aid in dying" can be morally justified. In a previous essay we examined a range of moral objections to euthanasia.[14] We argued that, on grounds principally of patient self-determination and relief of suffering, voluntary euthanasia for competent, terminally ill patients should be legalized, subject to adequate safeguards. Daniel Callahan has challenged the moral coherence of a policy of limited voluntary euthanasia in a recent essay.[15] We believe that it is important to address this challenge prior to considering the ethical significance of differences between PAS and PPE.

According to Callahan, both patient self-determination and relief of suffering push the logic of justifying euthanasia beyond the limits of legalized euthanasia that we endorse: the voluntary choice of competent, terminally ill patients. With respect to self-determination, why should euthanasia be limited to terminally ill patients whose deaths are imminent? Persons suffering from profound handicaps or in the early stages of Huntington's disease, AIDS, or Alzheimer's disease, who have years of life (and suffering) ahead of them, would seem to have just as much a right to end their lives by choosing PPE or PAS as a person like Timothy Quill's patient Diane, who faced imminent death from leukemia.[16] With respect to relief of suffering, why should euthanasia be limited to competent patients who have expressed a clear choice in favor of being killed? For the incompetent may also suffer profoundly from incurable conditions.

Callahan sums up his challenge as follows:

> Considered from these angles, there are no good moral reasons to limit euthanasia once the principle of taking life for that purpose has been legitimated. If we really believe in self-determination, then any competent person should have a right to be killed by a doctor for any reason that suits him. If we believe in relief of suffering, then it seems cruel and capricious to deny it to the incompetent. There is, in short, no reasonable or logical stopping point once the turn has been made down the road to euthanasia.[17]

Is the argument for limited euthanasia morally incoherent? Neither self-determination nor relief of suffering are absolute principles. Both are subject to practical limits necessary to accommodate them with other valid norms. We believe that there are good and mututally coherent reasons for limiting legalized PAS and PPE to voluntary requests of competent, terminally ill patients. The requirements of competence and voluntariness reflect the principle that persons have a right not to be killed, which only they can waive by their competent and voluntary choice to seek control and aid in dying. The suffering of the incompetent is a matter of grave concern. Under current law and practice, they are allowed to die by forgoing life-sustaining treatment with comfort care, authorized by their advance directives or the decisions of family or other surrogates. Nonvoluntary euthanasia goes too far down the slippery slope leading to involuntary killing of patients who are considered to be no longer worthy of treatment or care.

The requirement of terminal illness, which we define as a competently determined prognosis of no more than six months to live, may seem more diffificult to justify. Certainly, there is an element of arbitrariness in the specified time period: why not three months or one year? Yet the limitation of euthanasia to terminal illness and imminent death responds to the medical context of the policy. Our concern here is not the general conditions under which suicide and mercy killing may be justified; rather, the policy we rec-ommend concerns the intervention of physicians in causing, or contributing to, the death of patients. Strict limits are ethically required where licensed healers are responsible for assisting suicide or administering lethal doses of medication. The medical exit from life should be open only for competent patients who are dying.

There is no escaping the moral tension involved in PAS and PPE: The medical goals of preservation of life and healing vie with relief of suffering and respect for patient self-determination. Expanding permissible euthanasia beyond conditions of terminal illness with imminent death constitutes too great a violation of physicians' professional commitments to preservation of life. In the case of terminally ill patients with a prognosis of no more than six months to live, health cannot be restored by medical treatment, and preserving life only staves off an imminent death. If competent patients who are not terminally ill, or likely to die within a relatively short period of time, are permitted to end their lives by PAS and PPE, then euthanasia goes beyond "aid in dying." PAS and PPE can be legitimate medical services only for dying patients who choose to be killed.

In sum, the limitation of permissible euthanasia to voluntary requests of competent, terminally ill patients respects the right of patients not to be killed, except by their autonomous choice; and it reasonably accommodates the tensions within the "internal morality" of medicine between relief of suffering and preservation of life. Accordingly, we reject the argument that limited euthanasia is morally or logically incoherent.

⨉Moral Accounting of PAS and PPE

Are there morally significant differences between PAS and voluntary, active euthanasia? If a terminally ill patient wishes to control the process of dying rather than endure a "natural" death, is it morally preferable (or less morally objectionable), other things being equal, for the patient to die by physician-assisted suicide than by the physician administering a lethal dose of medication? It has been argued recently that it is important to distinguish PAS and PPE, both conceptually and ethically, and that PPE is ethically more problematic than PAS.

In a significant statement on the care of dying patients authored by twelve distinguished physicians, ten of the twelve endorsed the position that "it is not immoral for a physician to assist in the rational suicide of a terminally ill person."[18] The authors distinguished PAS from PPE as follows: "In the case of suicide, the final act is performed by the patient, even when the physician provides indirect assistance in the form of information and means. By contrast, euthanasia requires the physician to perform a medical procedure that causes death directly. It is therefore even more controversial than assisted rational suicide."[19] Although the authors do not explicitly condemn all cases of physician-performed euthanasia, they point out that "Many physicians oppose euthanasia because they believe it to be outside the physician's role, and some fear that it may be subject to abuse."[20] The preference for PAS over PPE may have been reinforced by Timothy Quill's widely discussed account of helping a terminally ill patient commit suicide.[21]

In a recent essay entitled "The Morality of Physician-Assisted Suicide," Robert F. Weir carefully distinguished PAS from PPE.[22] Weir noted five differences:

1. *Agency*—The physician contributes to the death of the patient in PAS but is not causally responsible for killing the patient. In PPE the physician intervenes, at the request of the patient, to cause the patient's death.
2. *Capability*—The patient in the case of PPE, but not PAS, is likely to be incapable physically or psychologically of causing his or her own death.
3. *Method*—PAS usually proceeds by the physician prescribing a drug such as barbiturates and informing the patient of the required lethal dose, whereas in PPE the physician injects the patient with a lethal dose of medication.
4. *Certainty*—Since suicide is up to the patient, who must act effectively to cause his or her own death, the physician cannot be sure that the patient actually will commit suicide. The physician prepared to administer a lethal dose of medication is much more certain that the intended result will be achieved.
5. *Legal liability*—PAS, in contrast to PPE, is not a crime in all jurisdictions; and where it is illegal, the physician is much less likely to receive any penalty than in the case of PPE.

What is the ethical significance of these five differences? The difference in agency raises the issue of responsibility. From a moral perspective, is the physician who assists in the suicide of a terminally ill patient less responsible for that patient's death than if he or she performed active euthanasia?

If the patient's choice of death is truly voluntary, then we can see no difference in moral responsibility between PAS and PPE. In both instances the physician lends his or her knowledge and skill to achieving the patient's chosen aim of ending life rather than suffering a natural death. The patient's voluntary preference is essential for the ethical justifiability of both PAS and PPE. Accordingly, the physician is the agent of the patient in permissible cases of PAS and PPE; he or she intervenes at the request of the patient to help the patient die.

Is the capability of the patient to commit suicide an ethically relevant consideration in favor of PAS as opposed to PPE? To legitimize PAS (in some cases) but not PPE would seem to discriminate unfairly against those who are physically incapable of causing their own death. If a physically capable and a physically incapable patient both are terminally ill and prefer to be killed rather than to suffer a natural death, on what grounds could it be justifiable for the former, but not the latter, to receive a physician's help in dying?

By "psychological capability" to commit suicide we mean the willingness of a person to cause his or her own death. Psychological capability might be considered of genuine ethical significance. If the patient chooses to be killed but cannot muster the will to cause his or her own death, when physically capable of doing so, then perhaps the patient's choice is not genuinely voluntary. Is it true, however, that the psychologically incapable patient does not genuinely choose to die, when he or she requests a physician to perform active euthanasia rather than to provide the means to commit suicide? Some people may be unable to bring themselves to do what they will. They are of sound mind and know that they want to terminate their lives but feel incapable of bringing this about by themselves. For others, lack of psychological capability to commit suicide may indicate lack of true volition to be killed. If voluntariness is not in question, why should a dying patient be denied release from suffering and control over death, owing to timidity or squeamishness?

On the other hand, it is important to note that because a person requests help from a physician in committing suicide and intends to carry it out, it does not follow that this choice is genuinely voluntary. The patient's judgment may be clouded by depression; the patient may be ignorant of the prospects of relieving suffering by means of hospice care. The ethical issue suggested by capability is voluntariness. Unless it can be shown that voluntariness can be assured in the case of PAS but not PPE, and that psychological capability is a reliable indicator of voluntariness, then this difference does not count ethically.

At first glance it may seem that the method, as distinct from the agency, of causing death is ethically irrelevant. What difference can it make if the patient dies from ingesting barbiturates or from a lethal dose of potassium chloride? On closer inspection, however, it can be seen that there are two competing considerations raised by the method of euthanasia. On the one hand, in favor of PAS Howard Brody pointed out "the potentially therapeutic effect of both having the means to end one's life available, and having personal control over the time and setting of its use."[23] The dying patient who opts for suicide when living is no longer bearable, has the security that comes from knowing that the means of swift and painless death lie at his or her disposal. The patient who has a "contract" with his or her physician to perform active euthanasia at the patient's request may always harbor some residual doubt that the physician will come through as promised. On the other hand, there is a consideration of method that may be seen as potentially favoring PPE over PAS. For a lay person administering a supposed lethal dose of medicine there is some risk that the attempt at suicide will fail. A botched suicide may cause the patient to receive unwanted emergency treatment and to lose control over his or her life. This is less likely to occur in the case of PPE, which brings us to the next point of difference.

The greater certainty of causing death in PPE, as compared with PAS, is a function of the differential factors of agency and method. Curiously, what is more certain from the perspective of the physician may be less certain in the mind of the patient. As indicated above, the patient who requests PPE may be uncertain about achieving death at the desired time, because the means of causing death lies in the hands of the physician, not the patient. What may seem to be more ethically troubling about PPE in this respect is not, strictly speaking, its greater certainty from the perspective of the physician but the greater potential for undue influence by the physician over suffering, vulnerable patients. The presence of the physician intending to cause death, in response to the patient's request to be killed, may be perceived by the patient as pressure "to get it over with" or as an authoritative endorsement of choosing death. The patient may have less psychological leeway to draw back at the brink in the case of PPE, as compared with PAS. Thus this influential presence of the physician may be regarded as a potential interference with patient's self-determination.[24] Nevertheless, presence of the physician provides an opportunity to check the decision-making capacity of the patient and the voluntariness of the decision to die before administering a lethal dose of medication. The physician who assists with suicide by prescribing medication may have no contact with the patient immediately prior to the act of suicide.

Does the difference in legal liability have ethical significance? Assuming that physicians, as well as all members of a liberal-democratic society, have a *(prima facie)* moral obligation to obey the law, then the fact that PPE is unquestionably illegal and that PAS may not be illegal counts in favor of PAS. This legal difference, however, is irrelevant when the issue under consideration is whether PAS or PPE should become legitimate, legally authorized medical practices. From the perspective of public policy, rather than the conduct of individual physicians under the legal status quo, this difference between PAS and PPE does not ethically favor PAS over PPE.

There is an issue intimated by the statement of the twelve physicians considered above that has some potential bearing on the relative ethical merit of PAS and PPE. PPE poses a greater risk of abuse in the following respect: Legalized PPE opens the door to euthanizing incompetent patients who would be incapable of committing assisted suicide. We argue below that this greater risk of moving down the slippery slope can be prevented or minimized by instituting a range of safeguards designed to assure that PPE and PAS are restricted to the voluntary requests of competent, terminally ill patients.

The discussion of differences between PAS and PPE surfaced two ethical considerations that suggest the potential superiority of PPE over PAS: In PAS there exists a greater risk of botching the procedure aimed at causing death, thus making it a less effective means of realizing the autonomous choice to control the process of dying by a terminally ill patient; and legalizing PAS but not PPE unjustly discriminates against competent handicapped patients who are physically incapable of committing suicide. A third consideration is the risk that patients who choose PAS in a context in which PPE is illegal will be abandoned at the point of death. If PPE or mercy killing by a lay person amount to murder, then reasonable patients who choose PAS may want to assure that their physician or family members are not in their presence at the time of suicide, so that they are not implicated in the deed. This issue arose in Quill's account of the assisted suicide of his patient Diane. At the conclusion of his narrative, Quill mused about the problem of abandonment: "I wonder whether the image of Diane's final aloneness will persist in the minds of her family . . . I wonder whether Diane struggled in that last hour, and whether the Hemlock Society's way of death by suicide is the most benign. I wonder why Diane, who gave so much to so many of us, had to be alone for the last hour of her life."[25] To be sure, some patients may choose to be alone in order not to weaken their resolve to commit suicide, or for other reasons. But if their aloneness is caused by a desire to avoid legal repercussions for others, then the risk of abandonment counts as an ethical consideration against legitimizing PAS but not PPE.

In a more recent article, Quill, joined by Cassel and Meier, asserted that physicians ideally should be present at the time of assisted suicide, which should be legalized so that patients need not be abandoned out of fear of legal consequences.[26] Although the recommendation of physician presence at the time of assisted suicide is reasonable, it weakens the authors' argument that assisted suicide, but not voluntary active euthanasia, should be legalized. They contended that "In voluntary euthanasia, the physician both provides the means and carries out the final act, with greatly amplified power over the patient and an increased risk of error, coercion, or abuse."[27] This raises the consideration discussed above of undue influence by the physician in PPE. However, if the physician is to be present at the time of assisted suicide, then it is doubtful that there is a significant difference between PPE and PAS regarding the risk of error, coercion, or abuse in the case of competent patients who choose to die. If the very presence of the physician may pressure the patient, "to get it over with," what difference does it make if the method of ending life is self-administered or physician-administered? Moreover, what is the physician who witnesses assisted suicide of his or her patient to do if the suicide attempt appears to be failing? If the aim is to provide aid in dying to a competent, terminally ill patient who voluntarily opts for death, should not the attending physician be prepared to administer a lethal dose of medication in case the suicide attempt fails? (For the sake of argument, we ignore the relevance of any legal distinction between PAS and PPE.)

A moral accounting of PAS versus PPE yields different relative disadvantages of each mode of helping terminally ill patients control their dying. There are three counts against PPE:

1. Potentially greater threat to patient autonomy and voluntariness owing to physician presence and direct lethal intervention;
2. Lack of therapeutic benefit that comes from placing the power to control the occurrence of death in the patient's hands; and
3. Greater risk of moving down the slippery slope from voluntary to nonvoluntary euthanasia.

These points need to be balanced against the three counts against PAS considered above. In view of these contrasting ethical disadvantages of PPE and PAS, the case for the moral superiority of PAS is doubtful. If, however, both PAS and PPE were legalized, subject to suitable safeguards, then patient choice with respect to dying would be enhanced over both the status quo and the prospect of making PAS alone a legally and morally legitimate option. We support a carefully designed policy of legalized voluntary euthanasia in the case of competent, terminally ill patients, which includes both PAS and PPE. This inclusive policy would obviate the ethical problems of forced abandonment of the patient at the time of death and discrimination against competent, but handicapped terminally ill patients. We believe that procedural safeguards, which we explicate below, can minimize, if not prevent, the remaining problems associated particularly with PAS and PPE and other often-cited potential harms of legalizing either form of voluntary euthanasia.

The point of instituting procedural safeguards in the context of a policy of legalized euthanasia is to assure that physician-assisted suicide and physician-performed euthanasia are limited to ethically permissible cases. We believe that PAS and PPE are legitimate only in the case of competent, terminally ill patients (with a prognosis of no more than six months to live) who voluntarily request to be killed. Thus the safeguards discussed next are designed to assure competence, terminal illness, and voluntariness.

Safeguarding Against Abuse: Committee Review

The literature on euthanasia contains useful discussions of a variety of safeguards to prevent abuse.[28] We focus here on one pivotal requirement, which has not received the attention it deserves. We believe that all cases of PAS and PPE must be reviewed in advance by a specially constituted committee designed to assure the patient's competence, the voluntariness of the request, and the terminal condition from which the patient suffers. (A more detailed consideration of committee review and other safeguards appears in our essay, "The Case for Legalized Euthanasia."[29])

The rationale for prior committee review draws on an analogy with human subjects research. Faced with abuses in the protection of human subjects of biomedical research (the Tuskegee syphilis study, the Jewish Chronic Disease Hospital study, and so forth) the US federal government mandated prior review of all proposals of research involving human subjects in institutions that receive federal funding. The knowledge that clinical research had moved down the slippery slope to serious violation of the rights and welfare of human subjects motivated the development of detailed regulations for the conduct of research and mandatory prior committee review to assure compliance with the regulations. We recommend prior committee review of requests for euthanasia to protect vulnerable patients and to prevent abuses.

Euthanasia committees should consist of interdisciplinary panels of clinicians, ethicists, lawyers, and laypersons. Members of the euthanasia committee would need the expertise to assess competence, voluntariness, and the prognosis of the patient. Interviews with the patient and the physician would be conducted by committee members to assure that the patient genuinely wishes to end his or her life by PAS or PPE, that he or she is not suffering from treatable depression or some other psychological or medical condition that may interfere with rational judgment, and, furthermore, that the patient is aware of alternatives to euthanasia, such as hospice care. One or more physician members of the committee would need to review the patient's medical condition to confirm the diagnosis of terminal illness and the reasonableness of the estimated prognosis of no more than six months to live.

It is important to dispel some misconceptions to which the proposal for committee review is liable. Alexander Capron declared, "By directly involving an official body, appointed under state procedures, it would seem to sanction the taking of life, a result that opens the door to abuses by the state and at the very least makes every act of euthanasia a collectively imposed sentence of death on innocent patients, some of whom would never have actually requested euthanasia."[30] Whereas a euthanasia review committee might function in this capacity, it is by no means inherent in the policy or practice of committee review that it would impose a collective, state-sanctioned "sentence of death on innocent patients." Capron assumed that committee review would authorize nonvoluntary or involuntary euthanasia. If euthanasia is limited by law to voluntary requests of aid in dying by competent patients, then committee review and approval, designed to achieve the objectives indicated above, cannot be conceived reasonably as imposing a death sentence on an innocent patient.

In a thoughtful essay discussing protections against potential abuses of legalized voluntary euthanasia, Margaret Battin remarked, "Notice what is *not* recommended here as a protective device: the deliberations of a committee. These can only be deliberations about the content of the patient's choice, and I do not see that a committee decision on whether the patient may or may not end his or her life protects the quality of the patient's choice."[31] Why the deliberations of a euthanasia review committee could concern *only* the content of the patient's choice is not made clear. Battin's statement begs an important question at issue in the proposal for mandatory prior committee review. Our proposal explicitly confines committee review to assurance of competence, voluntariness, and an imminently terminal prognosis. Determinations of the quality of the patient's life, the unbearableness of the patient's suffering, or the wisdom of seeking PAS or PPE do not fall within the purview of committee review under our proposal. Committee approval, based on satisfying the three procedural conditions, gives patients and physicians the license to undertake PAS or PPE. It does not impose euthanasia, since a patient at any time may freely decide not to undertake PAS or PPE; and no physician would be legally or morally obligated to comply with a patient's request for euthanasia. Nor does the committee either approve or disapprove the reasons for voluntary euthanasia.

In this respect our proposal for prior committee review for euthanasia differs from research review by an Institutional

Review Board (IRB). The IRB determines the reasonableness of
the risk–benefit ratio for human subjects posed by a research
protocol, in addition to assuring the adequacy of the process of
obtaining informed consent. We reject committee review concerning the substantive reasonableness of requests for euthanasia; for this involves an authoritative body passing judgment on
the meaning and quality of the life of a patient who faces imminent death. Provided that the review committee has determined
that the three procedural conditions are satisfied, whether or not
to undertake PAS or PPE remains up to the decisions of an
informed and willing patient and a willing physician.

Patients seeking euthanasia and physicians prepared to help
them, as well as euthanasia advocates, may object to committee
review. If euthanasia is seen as included within "the right to die,"
why should patients have to satisfy a committee regarding their
competence and the voluntariness of their choice? The fact that
death is the intended, immediate consequence of euthanasia and
that physicians are involved as agents of death make this dimen-
sion of the right to die properly subject to regulation. A civilized
society can permit the killing of patients by, or with the assistance
of, physicians only if stringent safeguards are instituted to prevent abuse. As long as committee review is confined to the three
issues of competence, voluntariness, and terminal prognosis, this
requirement should not be unduly burdensome.

Mandatory prior committee review of PAS and PPE would
necessarily interfere with the privacy of the physician–patient
relationship. We believe that the need to safeguard against abuse
outweighs the infringement of privacy. However, procedures
should be designed to minimize the threat to privacy; e.g., con-
fidentiality must be scrupulously protected. The experimental
policy of committee review should be monitored and subjected to
evaluative research, which might suggest ways to reduce the intru-
siveness and formality of review consistent with the aim of preventing abuse. Dying patients who might be unwilling to submit
to committee review of their request to be killed would remain

free to choose a "natural" death by forgoing life-sustaining treatment and seeking only comfort care.

Physicians might object to euthanasia committees as interfering with professional discretion, as time-consuming, as indicative of a lack of faith in their professional integrity, and as disruptive of the physician–patient relationship. It is arguable that if physician-assisted suicide or physician-performed euthanasia become legal practices, then their use should be left to the private deliberations of physicians and patients. According to this perspective, safeguards against abuse should be built into professional standards for PAS and PPE. Quill, Cassel, and Meier presented six clinical criteria for physician-assisted suicide:[32]

1. The patient must suffer from an incurable condition;
2. This suffering must not be "the result of inadequate comfort care";
3. The request to die must initiate from the patient and reflect a clear, voluntary, and persistent preference;
4. The physician must assure that the patient's judgment is rational and not clouded by depression;
5. There must be a "meaningful," but not necessarily a long-standing, relationship between physician and patient; and
6. There must be consultation with another physician to assure that the preceding five criteria are satisfied.

We endorse these criteria with the exceptions argued above that they should apply to both PAS and PPE and that euthanasia should be limited to terminally ill patients. Are these clinical criteria adequate to safeguard against abuse?

As standards for the *good* doctor, these clinical criteria are exemplary. But what about those doctors whose integrity is not above reproach or who are less than thoroughly conscientious? Consultation with another physician could help guard against failure to follow the other five criteria; however, this depends once again on the integrity and conscientiousness of the consultant. The problem here is reminiscent of the early debate over regulation of clinical research in response to unacceptable abuses of research subjects. Henry Beecher argued that protection of human subjects of clinical research should rest primarily with the integrity and conscience of investigators.[33] Instead, public policy, rightly in our opinion, instituted mandatory prior review of human subjects research by an interdisciplinary committee.[34] We believe that prior committee review of requests for PAS and PPE, along the lines suggested, should be mandatory. The seriousness of error is too great to make committee review merely optional at the initiative of physicians in doubtful or difficult cases. Therefore, unlike the situation in the Netherlands, performance of euthanasia should not be placed entirely within the professional discretion of physicians. The process of slippage into nonvoluntary euthanasia, which seems to have occurred under the Dutch experiment, as well as other abuses, would be checked by mandatory prior committee review.[35]

Conclusion

We have argued that physician-assisted suicide is not morally superior to physician-performed euthanasia. Indeed, both forms of aid in dying have distinct moral assets and liabilities. Accordingly, we do not endorse the argument that PAS, but not PPE, should be legitimized. We prefer a policy of legalizing PAS and PPE for the voluntary requests of competent patients suffering from terminal illness, subject to a series of safeguards designed to assure that the conditions for permissible euthanasia have been achieved. The most important institutional safeguard, in our opinion, is mandatory prior committee review, which we discussed briefly. Any experiment in legalized euthanasia would require very careful design, official monitoring, and evaluative research, aimed at determining whether the experiment is working as planned and abuses are being prevented. It is time to move—

with caution, resolve, and watchfulness—from the familiar arena of philosophical debate over the ethics of euthanasia to the uncharted waters of experimenting with legalized, voluntary euthanasia in one or more jurisdictions of the United States.

Notes and References

[1]President's Commission for the Study of Ethical Problems in Medicine and Biomedical and Behavioral Research (1983) *Deciding to Forego Life-Sustaining Treatment,* Government Printing Office, Washington DC, 15–41.

[2]National Center for Health Statistics (1990) *Vital Statistics of the US, 1987,* vol. II, Mortality, part A. (Public Health Service), Washington, DC.

[3]R. J. Miller (1992) Hospice care as an alternative to euthanasia, *Law Med. Health Care* **20,** 127–132.

[4]R. F. Weir and L. Gostin (1990) Decisions to abate life-sustaining treatment for nonautonomous patients, *JAMA* **264,** 1846–1853.

[5]D. Orentlicher (1992) Advance medical directives, *JAMA* **267,** 949–953.

[6]J. Rachels (1975) Active and passive euthanasia, *New Engl. J. Med.* **292,** 78–80.

[7]T. L. Beauchamp and J. F. Childress (1989) *Principles of Biomedical Ethics,* 3rd Ed., Oxford University Press, New York, p. 147.

[8]The studies in Table 1 were selected from a series of studies on euthanasia and assisted suicide collected by the Roper Center for Public Opinion Research, PO Box 440, Storrs, CT 06268. The studies reported asked the same, or virtually the same, question about legalizing euthanasia.

[9]Louis Harris Poll (1988) Making Difficult Health Care Decisions (for Loran Commission) New York.

[10]F. R. Abrams (1988) Licensed Physician Questionnaire—Anonymous Poll—Withholding and Withdrawing Life-Sustaining Treatment: A Survey of Opinions and Experiences of Colorado Physicians, Center for Health Ethics and Policy, Boulder, CO.

[11]D. M. Gianelli, Washington state physicians divided on proposal to legalize euthanasia, *Am. Med. News,* April 1, 1991, p. 1.

¹²The National Hemlock Society, 1987 Survey of California Physicians Regarding Voluntary Active Euthanasia for the Terminally Ill, The Hemlock Society, Los Angeles, February 17, 1988; S. Helig (1988) The San Francisco Medical Society euthanasia survey: Results and analyses, *San Francisco Medicine* (May, 24–26, 34; and M. Overmyer (1991) National survey: Physicians' views on the right to die, *Physicians Manage.* **31,** 7, 40–45.

¹³T. E. Quill (1991) Death and dignity: A case of individualized decision making, *New Engl. J. Med.* **324,** 691–694; G. I. Benrubi (1992) Euthanasia—The need for procedural safeguards, *New Engl. J. Med.* **326,** 197–199; T. E. Quill, C. K. Cassel, and D. E. Meier (1992) Care of the hopelessly ill—Proposed clinical criteria for physician-assisted suicide, *New Engl. J. Med.* **327,** 1380–1384; and H. Brody (1992) Assisted death—A compassionate response to a medical failure, *New Engl. J. Med.* **327,** 1384–1388.

¹⁴F. G. Miller and J. C. Fletcher (1993) The case for legalized euthanasia, *Perspect. in Biol. Med.* **36,** 159-176.

¹⁵D. Callahan (1992) When self-determination runs amok, *Hastings Center Report,* March-April **22,** 52–55.

¹⁶T. E. Quill, *op. cit.*

¹⁷Callahan, *op. cit.,* p. 54.

¹⁸S. H. Wanzer et al. (1989) The physician's responsibility toward hopelessly ill patients: A second look, *New Engl. J. Med.* **320,** 844–849.

¹⁹*Ibid.,* p.848.

²⁰*Ibid.,* p. 849.

²¹T. E. Quill, *op. cit.*

²²R. F. Weir (1992) The morality of physician-assisted suicide, *Law Med. Health Care* **20,** 116-126.

²³H. Brody, *op. cit.,* p. 1386.

²⁴We owe this point to a conversation with Howard Brody. Brody elaborates on this in his article, Assisted death, *op. cit.,* p. 1386.

²⁵T. E. Quill, *op. cit.* p. 694.

²⁶T. E. Quill, C. K. Cassel, and D. E. Meier, *op. cit.*

²⁷*Ibid.,* p. 1381.

²⁸*See,* e.g., M. Battin (1992) Voluntary euthanasia and the risks of abuse: Can we learn anything from the Netherlands? *Law Med. Health Care* **20,** 133–143.

[29]F. G. Miller and J. C. Fletcher, *op. cit.*

[30]A. M. Capron (1992) Euthanasia in the Netherlands: American Observations, *Hastings Center Report,* March-April **22,** 33.

[31]M. Battin, *op. cit.*, p. 140.

[32]T. E. Quill, C. K. Cassel, and D. E. Meier, *op. cit.* pp. 1381,1382.

[33]H. K. Beecher (1966) Ethics and clinical research, *New Engl. J. Med.* **274,** 1354–1360; and (1970) *Research and the Individual,* Little, Brown, Boston, MA.

[34]J. C. Fletcher (1983) The Evolution of the Ethics of Informed Consent, in *Research Ethics* (K. Berg and K. E. Tranoy, eds.), Liss, NY, pp. 187–228.

[35]C. F. Gomez (1991) *Regulating Death,* Free, New York.

The Ethics
of Physician-Assisted
Suicide

David C. Thomasma

When Dr. Jack Kevorkian, who is called "Dr. Death," was acquitted for the second time of murder charges on July 21, 1992, he said: " . . . it is time for other doctors to join him and make physician-assisted suicide safe, effective and widely available."[1] Opponents of physician-assisted suicide think, instead, that it is time to stop him before the State of Michigan, where he resides and practices his "craft," becomes the suicide capital of the world. Essentially the court judged that physician-assisted suicide is not a crime in Michigan, even when the patients so helped are not terminally ill. Kevorkian had been charged with murder when he assisted Sherry Miller who had multiple sclerosis and Marjorie Wantz who was victimized by severe chronic pelvic pain. On October 23, 1991, Miller used his machine to take a lethal over-dose of drugs, whereas Wantz inhaled carbon monoxide. Earlier charges for assisting Janet Adkins, who suffered from early stages of Alzheimer's Disease, were dropped in December, 1990.

More recently Kevorkian helped Catherine Andreyev, 46, who had suffered from cancer for 6 years. She was single, had no children, and her parents were already dead. Kevorkian said:

"The aim of suicide is to end a life . . . I consider this [carbon monoxide inhalation] a well-tested, well-controlled, well-thought-out medical procedure. The aim of this is to terminate unbearable suffering. I've made progress because for one more human being, suffering is ended."[2] In opposition to his methods, Reinhard Priester said, "He seems to be on a personal crusade not just to legalize it [but] to legitimize this type of activity."[3]

After the Miller and Wantz assisted suicides, Circuit Court Judge David Breck asked Kevorkian to stop counseling suicide patients until doctors and lawyers could formulate procedures, but Kevorkian said it would be wrong to stop: "You're going to watch a person suffer in agony while somebody's debating?" he asked. "I will wait, but not if the case is extreme."[4] Kevorkian's view is that only trained specialists called obitiatrists should be permitted to assist in suicide,[5] and that this has always been the responsibility of the medical profession: "This is a medical service. It always was."[6] Kevorkian is also in favor of gaining permission to harvest the organs of persons whom physicians assist in suicide, but recognizes that society is not ready for such action yet. Considering the depth of disagreement with his actions among the populace and profession, this is an understatement.

The seventh and eighth women to die with Kevorkian's help were assisted just before Governor Engler signed a law banning assisted suicide for 15 months until a commission studies the issue. Mrs. Lawrence suffered from heart disease, emphysema, and arthritis and complained about her constant pain. Mrs. Tate suffered from amyotrophic lateral sclerosis, and could only communicate by typing on a keyboard. Once the legislature receives the report of the commission, it will have 6 months to draft appropriate legislation. Although the expected outcome appears to be neutral, the Governor said in signing the ban:

> I want to sign it today as a protest to what Mr. Kevorkian has done. The methods of Mr. Jack Kevorkian—and I stress "Mr." since his license to practice medicine has been sus-

pended—are wrong because he has deliberately flouted the law and taken it upon himself to be his own judge, jury and executioner in Michigan.[7]

Through his lawyer, Kevorkian responded that this ban will not stop him from helping others who suffer because "He is held to a higher standard."[8] A court trial is now scheduled under the new law for the 17th patient Kevorkian assisted.

I have spent some time with the case of Dr. Kevorkian, since in it are embodied all the values at risk in favoring and opposing physician-assisted suicide. The legal, political, and moral problems with physician-assisted suicide are many. First, the legislature in the State of Michigan enacted a law to make physician-assisted suicide illegal, as it is in other states. Otherwise, the courts could not rule against Kevorkian's actions. Despite what the Governor said, Dr. Kevorkian originally broke no law. Note how the Governor's response to Kevorkian mentions how the latter took upon himself the role of judge, jury, and executioner. In other words, the strict guidelines that might someday be applied to permitting killing in any form, as they have been developed over the centuries for permitting killing in self-defense or in executions, have not yet been developed for physician-assisted suicide in the United States.

Since the legal picture is still unclear, it is obvious that assertion and counter-assertion will prevail for some time. Americans tend to debate issues like physician-assisted suicide in the midst of efforts to legislate an outcome. Simultaneous with the public debate, powerful political forces on both sides of an issue lobby for their own point of view. Usually this point of view does not take sufficient notice of the merit of the opposing side. Political pressure and not reasoned compromise overwhelms the issue.

But can one compromise on fundamental moral issues, particularly those that affect the value of human life and the standards by which we honor that life in others? Beyond the legal and political issue, then, lurks the difficult moral question about the taking of human life under any circumstance. Furthermore, is it true that assisting the suicide or death of a patient has always been, as Kevorkian claims, and other physicians like Timothy Quill, Christine Cassel, and Howard Brody now support, the responsibility of organized medicine? Is there such a thing as rational suicide, even when one is not yet dying? Can individuals request assistance from others for dominion over their lives? Are our lives our own, under the command of autonomy, or are they gifts from a Higher Power to whom alone matters of life and death must be left? What about the ethics of intervening, using our intellectual and moral capacities, in the lives we are given, even if they come from God? Is not taking responsibility for our own technology not part of the gift of freedom and control we have been given, granting a theological perspective to human life?

At stake, then, in the public and highly politicized debate over physician-assisted suicide are some fundamental questions to be addressed here:

1. Can and should individuals have control over their own lives, even to the point of ending them for a reasonable purpose?
2. Does or should the doctor–patient relationship include a responsibility to assist individuals who wish to end their lives for either reasonable or unreasonable purposes?
3. Should society permit physician-assisted suicide and, if so, under what conditions? Would these conditions protect against indiscriminacy in a violent society?
4. What about the rule against killing? Can it be suspended, as it is under the strict guidelines affecting just war theory or capital punishment?
5. Is there a morally valid, if not legally valid, distinction between active, direct euthanasia, passive, indirect euthanasia, and physician-assisted suicide? Particularly, does intention play a role in the distinctions, if there are to be

any made? Furthermore, what difference would intention make anyway for the moral character of individuals and society?

The rest of this chapter will review the literature and address the issues in each one of the five categories mentioned. Before concluding, I will summarize what I consider the strongest arguments on either side of the issue. It would be presumptuous to suggest that the subsequent treatment would be comprehensive, given the far-ranging nature of the debate and legislative proposals already present in the United States and worldwide. Instead I focus on the questions formulated as touching on the heart of the debate about physician-assisted suicide.

Autonomy and the Control of One's Life

The first question about physician-assisted suicide has nothing to do with the physician, but everything to do with the patient. Can and should individuals have control over their own lives to such an extent that suicide be morally permitted?[9,10] At one time, suicide was proscribed by all our states. The origin of this proscription was both religious and secular. The religious basis of the laws against suicide arose from the belief that life was a gift of God, and that only God could take it back. Persons who committed suicide were not permitted a burial in the church cemetery, but rather were buried at a crossroads. The latter represented a widespread superstition that persons who committed suicide were "troubled," and that if the devil were passing over the gravesite to claim that troubled soul, he would see the cross in the crossroads and leave the person alone. The secular basis of a proscription against suicide in Anglo-Saxon law seems to have arisen from the chronic population shortage England experienced throughout its historical struggle with the powers on the continent. To commit suicide, then, was to deprive the King of the able-bodied person to enlist in the war against the hated French!

In a modern, secular society, these reasons no longer carry weight. Sufficient understanding of the motives and circumstances surrounding suicide exist such that persons who do commit suicide are still given religious burials today, and the laws against suicide have been stricken from the books. Nonetheless, some moral opprobrium, and its attendant embarrassment for the family, still exists about the practice. People still whisper about a suicide death. Attempted suicide is still treated as caused by depression or some other mental disorder, and all our interventionist efforts are oriented to getting help for such people. This remains a wise course, since many suicide attempts are treated as cries for help. Some of the criticism of Dr. Kevorkian's assistance lies in the fact that he did not know the individuals who requested his help, and that insufficient efforts were made to help the persons who seemed to the casual reader simply to be depressed about their illness that often was not yet life-threatening.

Once the religious conviction about the source of life and death has eroded in society, we are left with the view that each individual controls his or her own life.[11] This autonomy-based view of human life places the burden of continued existence squarely on the shoulders of the individual, rather than society. Given the sometimes horrible circumstances of human life, remarkably few people choose suicide as a way out.[12] Increasingly, however, long, debilitating, chronic illnesses lead some people to chose to end their lives.[13] This is done in the context of care, in a nursing home or as a member of a family, that supports the individual. When this happens the caregivers often think that they were responsible, somehow, since their care obviously was judged insufficient to surmount the growing depression of their loved one or patient. But this is not usually true.

Instead, the individual makes a rational judgment that his or her life is no longer worth living, given the burdens of illness, economics, or social rejection. In his 80s, increasingly upset over his growing incapacities and public criticism of his foundational work with autistic children, especially claims that he had been abusive, Bruno Bettelheim took his own life by tying a plastic bag over his head. For some this would represent a failure of the proper kind of care. For others it represents the triumph of the human spirit over terrible circumstances. He had outlived his life and his work, and wanted to die with some dignity.

Some wish to argue that rational suicide does not exist, not even in old age. Erikson argued that by aging we advance toward ego integrity. It is "the acceptance of one's one and only life cycle as something that had to be and that, by necessity, permitted of no substitutions . . . the possessor of integrity is ready to defend the dignity of his own life style against all physical and economic threats . . . In such final consolidation, death loses its sting."[14] Erikson further argued that in Western society we do not respect the fullness of the life cycle in order to help persons bring about this ego-integrity. But as Harry Moody asked, reflecting on Bettelheim's suicide, cannot ego-integrity also lead to a separate choice, one's choice to put a closure on one's life as a result of old age, and not of terminal illness? As Moody further reflected, "rational suicide by the very old expresses an attitude toward the past: it expresses a feeling of completeness as judged for oneself. The judgment is, simply, no more of life is worth living."[15] Moody also noted that such suicide is less "rational" than it first appears; actually it is based on a loss of faith in what life may still have in store for us. Perhaps what is needed, he noted, is a different kind of faith in life and in the community of caregivers.

The conflict between intervening to convince an individual of the worthiness of life and of respecting a person's judgment on his or her own quality of life makes suicide a very difficult ethical topic. Too often the community withdraws from a person who struggles to create meaning in his or her life, and that person then decides that life is no longer worth living. Only at that point, then, the community finally comes to his or her aid through emergency services. At the Society for Bioethics Consultation's annual meeting in Chicago, 1992, clinical ethics consultants debated a case in which withdrawal of care was contemplated for a now-incompetent patient who had attempted suicide. Such a case raises all the questions so far addressed: Is the suicide attempt a rational act of self-determination? Or does it stem from pathological causes? What is the extent of our duty to a person who attempts suicide? Should we intervene to save the life? Or should we instead honor that person's autonomous decision?

Society does play a role in suicide as well. When the Supreme Court in the Cruzan case decided that families do not automatically have the right to speak as surrogates for their loved ones, Dr. Christine Cassell spoke about the concerns of her elderly patients that their wishes would not be honored when they became incompetent, and that they would rather commit suicide than be put into hospitals and nursing homes where they would be stripped of their values.[16] The case of Jean Elbaum in New York State illustrates the tragic consequences of ignoring patient wishes. In 1986 Jean Elbaum was admitted to Grace Plaza Nursing Home after a stroke. From that time on she was incompetent. In October of 1987, her husband, Murray, requested that fluids and nutrition be withdrawn based on her prior choices. Since New York is one of the few states that requires "clear and convincing evidence" as the legal standard for withdrawal of fluids and nutrition, the nursing home refused to comply, arguing that no adequate evidence of her wishes existed. Mr. Elbaum countered by refusing to pay for the care. After two levels of court review the New York State Appellate Court, 2nd Division, ultimate concluded that sufficient evidence of Jean Elbaum's wishes did exist to meet the strict evidentiary standard, and it ordered the nursing home to comply with Mr. Elbaum's request or transfer Jean to a facility that would. Such a transfer occurred, and she died in August, 1989.

However, the case continued, since a dispute arose about paying for the $100,000 cost of unwanted care. After two more lower court decisions, the same Appellate Court made a shocking decision that Mr. Elbaum would be responsible for the costs of the care, since, at the time of the initial request made by Mr. Elbaum to withdraw the care, a court had not determined the adequacy of evidence about Mrs. Elbaum's wishes. Connie Zuckerman, an ethicist and lawyer, commenting about this case, noted that it establishes a terrible precedent that would lead most nursing homes to ignore family requests based on patient preferences, since no financial harm will come from this action, and that all such decisions, even ones in which all caregivers agree, would need judicial review before being implemented. Most especially, Zuckerman opined, health care facilities with all their expertise we would not seem to carry any obligation to work with families to resolve this kind of dispute.[17] Zuckerman noted that the New York State Task Force on Life and the Law will consider the technical problems in cases like the Elbaum case, but probably will not address what she calls "an almost exuberant arrogance" on the part of the nursing home.[18] Mrs. Elbaum had never "chosen" the nursing home or its care; she had been transferred there after her stroke.

Such attitudes will almost certainly lead to increasing suicide among the elderly, along the lines suggested by Dr. Cassell, and for the reasons she expounded. Kevorkian's willingness to consider requests from individuals who are not yet dying, or in the terminal phase of an illness, stems in part from a straight-on humanist view that individuals can control the circumstances of their own life, and from a perception that, when this control is in jeopardy, such persons have a right to eliminate the suffering a loss of control creates by committing suicide with the help of a physician.

Any public policy about physician-assisted suicide, then, must take into account the public and private moral ambivalence about suicide. At the very least, rational suicide must be distinguished as far as possible from that prompted by inordinate depression or other mental pathology. Specific conditions and responses to those disorders must be present. Further, every effort must be made by society, through its laws and procedures, to protect

individual values as far as possible, and to shore up family surrogate decision making in areas where no advance directives have
been made.

Physician-Assisted Suicide
in the Doctor–Patient Relationship

The most important feature of the debate about bioethics in
the United States today is the growing awareness of the inad-
equacy of the autonomy assumption.[19,20] Increasingly, as bioethicists in the United States encounter their colleagues from other
parts of the world, the autonomy assumption becomes more glar-
ing as a component of our thought. I first encountered this
assumption in a 1984 conference of Dutch colleagues who formed
the nucleus of the Dutch philosophy of medicine society. The
purpose of the conference was to critique a new book by Pellegrino
and me on the philosophy of medicine.[21] During the conference,
Dutch professors continually critiqued the assumption of indi-
vidualism, individual rights, and the primacy of the individual
patient in the doctor–patient relationship. Exasperated, at one
point I made the claim that everyone would agree that the notion
of individual autonomy was the most revolutionary in all
of human history, for it set limits on the state and the community
to govern the behavior of individuals. None of the forty or so
professionals attending the conference in the Netherlands agreed.
It was a cultural shock to learn that they considered the most
revolutionary concept in human history to be socialism. I later
returned to the United States and wrote about this difference.[22–26]

Since then the debate about the role of autonomy in the
doctor–patient relationship has centered on the proposal by H.
Tristram Engelhardt, Jr., that the basis of all bioethics, of all
ethics in fact, is respect for autonomy. Engelhardt's argument is
that it is impossible to be ethical if one ignores the individual's
autonomy, preferences, values. For Engelhardt, such values are

supreme in all decision making.[27] This proposal has implications for every aspect of the philosophy of medicine. In it, the individual reigns supreme, not the community. I do not have the space here to critique the autonomy assumption as it deserves. Lately, a movement called "communitarianism" has sprung up, led by professor A. Etzioni at Washington University in St. Louis. This group stresses that with powerful individual human rights come powerful human responsibilities for meeting the community's needs.

Nonetheless, the importance of the autonomy assumption for physician-assisted suicide is essential. In the past, the view of the doctor–patient relationship was largely dominated by the Hippocratic ethic, one which stressed the knowledge of the physician, and frank paternalism in relation to the patient. In fact, Hippocratism explicitly ruled out abortion, euthanasia, and physician-assisted suicide at a time when such actions were quite popular among other kinds of physicians. The Hippocratic oath should be seen as an effort of reformed-minded physicians, a minority and revolutionary group, to determine how medicine should be properly practiced. Therefore, as Cameron argued, Hippocratic medicine rejected pre-Hippocratic ethics based on relief-of-suffering, and replaced it with a sanctity of life ethic.[28]

Today's physicians understand their role as a balance between relief of suffering and nonharm.[29] With the rise of the autonomy ethic a counterbalance to the Hippocratic view has arisen. As an additional consequence, the internal morality of the profession began to be questioned. This can be seen from a report of a meeting between Dr. Kevorkian and the Michigan State Medical Society. Kevorkian's ideas received little credence from the assembled physicians, including the notion that since certain procedures for assisting in dying are clearly medical, and because they have been performed for centuries by people in power, they can be considered ethical. Yet Howard Brody, who chairs that committee, said that Kevorkian "impressed many people in the room by his presentation."[30]

In fact, the actions of Timothy Quill who assisted the suicide of one of his dying cancer patients were much better received in the literature than those of Kevorkian. He did so with a patient, Diane, who was clearly going to die, since she chose no longer to pursue the possibility of a 25% chance to live. Furthermore, the patient and he knew one another for years. In addition, the patient continually petitioned him for help on the basis of her values, values that he himself knew were permanently part of her identity. He tried for some time to persuade her toward other possible ways of coping with her terminal illness. In the end he decided to help her, although he remains concerned to this day about the way he was not present, nor were any of her family present, during her death, since legally they might have been "accomplices" in her suicide and subject to penalties.[31] Since then, Dr. Quill has noted that his actions were not unique. Many physicians have aided their patients in this way. He was unique in writing and speaking about it.

Indeed, a great deal of courage was needed to begin to discuss physician-assisted suicide and active, direct euthanasia, since the legal climate in the United States and elsewhere does not lend itself to this exposure. Recently a physician in England, Nigel Cox, was convicted of attempted murder of Mrs. Lillian Boyes, aged 70, who suffered intractable pain from rheumatoid arthritis not relieved by morphine or heroin. She had requested a fatal injection, and he had refused, just as Dr. Quill did earlier for his patient. Later, however, he gave her a lethal dose of potassium chloride by intravenous injection. He was convicted of attempted murder because the body had been cremated, and no final cause of death could be proved.[32] Stanley Rosenblatt recounted the case of Patricia Rosier, who sought the help of her prominent physician husband, Peter, to end her cancer-ravaged life with some measure of dignity. Peter did so, was indicted for first-degree murder, and now faces death in Florida's electric chair.[33]

This case and Dr. Quill's demonstrate the fundamental difference between Kevorkian's methods and Quill's. For sure, Kevorkian does not know the patients as well, or try to persuade them much not to take their own lives, and he does seem to want to gain publicity or notoriety in a way that Dr. Quill does not. Yet the fundamental difference lies in their respect for autonomy. Kevorkian believes that individuals have an absolute right to control their own lives and, when they judge them to be "over," to take their lives as a way out. In fact, this may be the basis of his view that he is responsible to a "higher law" than that of civil society, for in his view, the goal of the medical profession is to relieve suffering at the patient's request. Dr. Quill seems to be operating from an older version of the doctor–patient relationship, one in which patient autonomy is circumscribed by physician duties to preserve and prolong life. His sensitivity to these values has been praised in my presence often. Only when he saw that his patient would not choose these values any longer and would find some way to end her life, did he agree to help her for compassionate reasons.

Yet a question remains, perhaps the question that would impress many on the Michigan State Medical Society committee. In the end, if patient autonomy prevails over the values of the medical profession to preserve life, what is the real difference between the actions of Kevorkian and Quill? It seems only procedural differences lie between them, and if this is so, the real difference of opinion about physician-assisted suicide and euthanasia among supporters and objectors lies in the nature of the doctor–patient relationship itself, and in the limits and powers of that relationship.

Leon Kass' argument that the internal morality of the profession requires that doctors "give no deadly drug" has a bearing on the debate at this point. He argued that at the point we press for active euthanasia and physician-assisted suicide, we should instead accept the opportunity to learn the limits of medicalization of life and death, and instead learn to live with and against mortality. He held that there is a residual human wholeness in dying, however precarious, and that wholeness should be cared for even in the face of incurable illness and dying. His view is the opposite of both Kevorkian and Quill, and the many others who support them. He thought that, should doctors abandon their posts during the dying process in favor of "technical dispensing of death," they would set the worst example for the community at large.[34]

This reflection turns naturally to a consideration of social regulation of the profession and of the action of physician-assisted suicide.

Social Regulation

Absolutization of the patient's autonomy is a subject of growing tension. Concerns about libertarian assumptions implied by this emphasis have led many thinkers to counter autonomy with the need for beneficence as well.[35,36] The implications of conflicts about medical ethics and ethical theory for active euthanasia include the libertarian push for active euthanasia that endangers the health provider's values in caring for the dying patient. This push may diminish the moral quality of the relation between physician and patient. It clearly tends to place exclusive emphasis on the needs and wants of the individual patient. Ultimately euthanasia raises questions about the kind of society we ought to be.

There is serious concern about the impact on the community when physicians are involved in voluntary active euthanasia.[37] Leon Kass presented in another article a thoughtful articulation of what is owed a dying patient by the physician. He argued that humanity is owed humanity, not just "humaneness" (i.e., being merciful by killing the patient). Kass argued that the very reason we are compelled to put animals out of their misery is that they are *not* human and thus demand from us some measure of humaneness. By contrast, human beings demand from us our humanity itself. This thesis, in turn, rests on the relationship "between the healer and the ill" as constituted, essentially, "even if only tacitly, around the desire of both to promote the wholeness of the one who is ailing."[38] This is still a majority view among physicians.

Studies have shown that physicians do not evaluate whether a patient is dying solely on the basis of biomedical data. They also take into account the important features of human interaction, and therapeutically available interventions.[39] Such interactive concerns tend to present counterpressures to a straightforward honoring of patient wishes and autonomy with respect to euthanasia requests. Needless to say, fears about litigation might also contribute to reluctance to honor patient requests, and even for increases in pain medication.

The temptation to employ technology rather than giving oneself as a person in the process of healing is really a "technological fix." The technological fix is much easier to conceptualize and implement than the more difficult processes of a truly human engagement. The training and skills of modern health professionals overwhelmingly foster the use of technological fixes. By instinct and proclivity, all persons in a modern civilization are tempted by technical rather than personal solutions to problems.

As I noted above, medical technology gives us enormous power at all levels of life, but especially at the end of life. Yet concerns should not be confined to dispatching persons too early by injections in active, direct euthanasia, while not meeting their physical and social needs. Another form of the "technofix" society is to prolong suffering in conditions of hopeless injury to life. "Hopeless injury" as Braithwaite and I defined it, is:

> a condition in which there is no potential for growth or repair, no observable pleasure or happiness from living . . . and a total absence of one or more of the following attributes of quality of life: cognition or recognition, motor activity, memory or awareness of time, consciousness, and language or other intelligent means of communicating thoughts or wishes.[40]

Daily life is full of interactions with "things"—nonhuman
and fundamentally incomprehensible to most persons. We sometimes get so used to technological processes that we behave as
though they are substitutes for human and compassionate care.
Eating for many elderly and dying patients has been replaced by
tubes; participating in the spiritual and material values of human
life has been replaced by "merely surviving," as a being subju-
gated to the very products of human imagination. As Illich observed:

> Medical civilization is planned and organized to kill pain, to
> eliminate sickness, and to abolish the need for acts of suffer-
> ing and dying . . . [41]

> The new experience that has replaced dignified suffering is
> artificially prolonged, opaque, depersonalized maintenance.[42]

Such "beings" on depersonalized maintenance may no longer
be as human as the rest of us, precisely because of this subjugation. This is no way to respect the value of human life. Is a permanently unconscious being without any ability to relate to its
environment a "person"? Part of taking responsibility for our
technology is to avoid this subjugation of human life to machin-
ery in the first place, through more thorough discussions of possible outcomes and patient values regarding them.

With respect to the role of autonomy in the debate, then, the
greatest danger is to ignore lessons learned from world history,
as well as to make assumptions about the use of similar terms in
totally different cultural contexts. Some of the factors shaping
today's debate are very close to those that shaped the discussion
in Nazi Germany that led to social programs to eliminate the
vulnerable and weak.[43] The movement toward legalizing euthanasia was widely discussed, as it is in the United States today. It
was from the arguments of mercy that Hitler signed into law the
permission for certain designated physicians to kill patients

(Gnadentod) judged "incurably sick by medical examination."[44] This order is not far from the suggestion of Dr. Jack Kevorkian today that only certain, trained physicians, called obitiarists, would be established to carry out euthanasia for those who request it.[45]

Hitler's order soon was focused on the retarded and mentally ill, however. By 1941, 70,000 patients in mental institutions had been euthanized, paralleling similar mercy killing in other countries.[46,47] For the Nazis, this was just another step toward greater social hygiene, the elimination of gypsies, Jews, and socialists of all stripes. It takes a powerful state to unleash these destructive forces within the medical community itself. Yet the power of the state, rather than the empowerment of individuals with respect to euthanasia, was the cultural background.

This makes all the difference about the meaning of the concept itself. At first blush, explicit downgrading of the intrinsic value of human life seems very foreign to our way of thinking. Most Americans would react with dismay, even anger, at such a statement. We also have the advantage of hindsight about the horrible, devastating consequences of such thinking. The stench of the death camps is a pall that still hangs over Western Civilization.

Yet attitudes of superiority and of disvaluing individual lives creep into our own thinking as well. They creep in while constructing what we consider to be rational allocation plans for health care. Although we intend to minimize suffering and maximize the common good, individual persons in high numbers face neglect of their needs. Perhaps the biggest worry about euthanasia today is the forthcoming crisis in health care that will be created by an increasingly elderly population. Those over 85 years of age in the next 50 years will increase fivefold, from three million to 15 million citizens. Many of these persons will be dependent on long-term nursing home care. Such care is the most expensive medical cost for the elderly. In all state budgets, Medicaid is the second largest budget item after education. Persons over 85, in fact, take four times as much money to cover a hospitalization than those under 85. There will be fewer individuals "in the middle," able to bear the burden of caring for the young and the elderly. Already the phenomenon of the elderly (70–85 years of age) caring for the "old old" (those over 85) has begun. How soon would we turn our attention to the high cost of caring for extremely elderly and debilitated persons? Already the discussion of euthanasizing the demented elderly has begun again in the United States.[48,49]

Recall that the fundamental argument for forcible euthanasia is always economic. Once the activity of active voluntary euthanasia is in place, habits of finding other areas for "mercy," coupled with hard economic times, could easily lead to involuntary euthanasia. This is yet another form of the traditional slippery slope argument. It is based on fears about the violence in American society and the natural human propensity to find technical solutions to difficult social problems, ones we can only imagine for the future, but ones that our children must face soon enough. If we set the wrong precedents in the present, why would we think the same sort of thinking that occurred in Nazi Germany would not arise again?

That being said, it is also important to recognize that a constant focus on the voluntariness of euthanasia in all discussion and debate about euthanasia can assure avoidance of mercy killing. What the Nazis did was not euthanasia, but murder, purely and simply, for the good of the state. If quality of life judgments and social utility judgments are constantly held in check by our cultural and historical memory of the Holocaust, then we will be well on the way toward providing a humane solution to the problem of suffering not only at the end of life, but throughout it as well. We must find a compassionate and just vision of equality in our health care system today before there is any implementation of physician-assisted suicide and euthanasia.

The Rule Against Killing

The broader debate about active euthanasia in society today exposes contradictions in our attitudes and behavior about the "Rule Against Killing." This ancient rule is embedded in Western Civilization, such that only strictly controlled killing is permitted, that is killing out of self-defense, in military engagements, and in capital punishment. The strict control has to do with the moral rules by which the Rule Against Killing is temporarily and on a case-by-case basis suspended. For example, the Just War Theory was developed to detail the particular circumstances under which a person could suspend ordinary rules against killing. Why does society, however, permit killing in self-defense and defending the rights of others in a "just war" or in capital punishment, but not permit killing out of mercy, or participating in the rational suicide of a patient by providing the means?

One answer to that question must come from the backdrop of the Nazi mercy killing program. There is serious and accurate concern that today's physicians will very easily "tilt" in the direction of mercy killing, the same way the Nazi physicians did so. This objection to the suspension of the Rule Against Killing represents a conflation of all the desolate reflection on how physicians in Germany could have been so supportive of Nazi initiatives, especially those of "biological purity." In a volume devoted to the Nuremberg Code, Elie Wiesel agonized: "That doctors parti-cipated in the planning, execution, and justification of the concentration camp massacres is bad enough, but it went beyond medicine. Like the cancer of immorality, it spread into every area of spiritual, cultural, intellectual endeavor. Thus, the meaning of what happened transcended its own immediate limits."[50] Wiesel wrote of a famous Jewish professor, Shimon Dubnow, whose own student, Johann Siebert, not only taunted him in the Ghetto, but also eventually killed him. Wiesel wondered: "I couldn't understand these men who had, after all, studied for 8, 10, 12, or

14 years in German universities, which then were the best on the Continent, if not in the world. Why did their education not shield them from evil? This question haunted me."[51] The editors, Annas and Grodin, asked explicitly, "How could physician healers turn into murderers? This is among the most profound questions in medical ethics."[52]

The best answer to the question involves many different and complex factors. Society, itself, was primed to develop a biological basis for its political platforms. The use of the best of the new science of genetics by the Nazis is well known. What is not as well known is that over half of all practicing physicians joined the Nazi party early on, even before Hitler came to power. The sad record is that many more than the forty-some physicians prosecuted in the Nuremberg trials participated in planning and carrying out the various programs that now have become so infamous. What is worse is that most of them, such as the euthanasia program, were justified by international practices, particularly by laws and procedures in the United States.[53] Part of the justification for sterilizing the retarded was to clean up the genes of the rest of the race, and part of the reason for euthanizing the demented was economic, a "preemptive triage," to free up beds needed for soldiers in the war effort.[54] But both of these initiatives against "worthless life," were based on published papers around the world in which similar proposals were being made, e.g., in the *Journal of the American Psychiatric Association,* where killing the retarded, "nature's mistakes," was advocated.[55]

The lessons to be learned from this experience are that each individual must be treated as an end-in-him- or herself, that the principles of wartime triage should not become ordinary or accepted ethical practices, and that a desire to practice modern, genetic-based health care will inevitably lead to efforts to "keep up" with the world literature and standards of care elsewhere. Nazi physicians worried a lot about how the US genetic laws were more advanced than theirs. A final point is this: Nazi physicians did not lose their sense of right and wrong. Their perception of the good was colored by their society's mores, and the state of their own craft at the time. The leading Nazi medical ethicist, Rudolf Ramm, echoing an earlier sentiment, said in 1942, "Only a good person can be a good physician."[56]

History can easily repeat itself. The discussion of euthanasia and physician-assisted suicide escalates in the United States, and the practices of abortion and euthanasia spread around the world. When this is coupled with increased attention to genetic therapies, what is the "good" that will infuse the virtues in medical practice? Rational discussion in academic literature is not enough to provide the proper checks and balances on physicians in modern society. One person's good is another's evil. Training in the virtues, as Wiesel noted above, does not guarantee a good outcome; the mores and standards of society in conjunction with that training are essential. This is the reason that the virtues in medical practice must be coupled with a principle-based ethics. Further, neither one alone, nor both conjoined guarantee good behavior. Only critically reflective medical ethics and self-critical individuals of good character can offer some hope that history will not be repeated here. Science and medicine do not just serve external interests. They are also informed by and give credence to those interests. My claim here is only that a person of integrity would be less likely to succumb to the fancies and foibles of any particular era.

Intentionality and Relevant Distinctions

In 1991, the Dutch issued a report on the status of euthanasia called the Remmelink Report. Among other data in the report was the fact that about 8% of terminally ill patients in the Netherlands request direct euthanasia, 80% of these being performed at home by their general practitioner. About 4000 Dutch citizens die this way each year, 85% of whom are suffering from terminal cancer. Dr. Pieter Admiraal, president of the Dutch Medical Society and an early champion of euthanasia, noted that "Active euthanasia can only be the last dignified act of terminal care."[57]

Following this report, there was intense criticism of a category in which cases of "involuntary euthanasia" were summarized. For example, a representative of the American Bishops' pro-life secretariat last year claimed that the Dutch government itself had documented 1000 cases of involuntary euthanasia, in which "physicians had killed patients without their explicit request."[58] This charge is now echoed by many thinkers, some of whom are bioethicists and philosophers of medicine, in the United States and abroad, including the Netherlands itself. To the American lay and professional public this charge sounds like the Dutch are experimenting with Nazi mercy killing.[59]

Nonetheless, the charge can be strongly disputed and has been by those skilled at caring for the hopelessly ill in the clinical setting. Dr. Admiraal himself disputed the notion that Dutch physicians are killing unwilling and unwitting patients, noting that there has been only one such case over the past 20 years, and that physician was immediately and severely punished.[60] The data should, instead, be read in light of the clinical experience noted above. First and foremost, the Dutch medical system provides medical treatment for all patients during their dying process. There is no pressure to terminate one's life on the basis of economic concerns. Second, the kinds of cases listed in the "involuntary" category in the Remmelink Report are ones that occur in medicine around the world every day. For example, a defective newborn infant might be denied treatment by physicians because its life would be completely senseless. Courts have backed up the decision of physicians not to offer medically futile treatment in the Netherlands. Cases like this are reported as "involuntary euthanasia" because the Dutch refuse to draw the distinction between active and passive euthanasia, holding that the distinction has no moral merit. There is no difference in the intention that the patient achieve relief of suffering through death. Another example of a case in this "involuntary euthanasia" category would be a patient suffering terminal cancer of the head and neck that has invaded the carotid artery. Twice a shunt has been tried, but eventually the physician can predict that the patient will die of bleeding in the neck and brain. A nurse summons him when this eventuality occurs. He gives an order for an increased dose of morphine so that the patient will die seconds before the terminal bleeding takes her life. Lack of clinical insights can severely impair the "translation" of data from one context to another.

These realities contribute to the problem of maintaining "purity of intention." For many years, now, ethicists have argued that there is no moral distinction between killing and letting die. Yet there is a distinction from the point of view of intention. If one intends to kill directly by one's action, and one's actions are meant to bring about that death as a means of relieving suffering, then withholding or withdrawing care or actively killing the patient makes no moral difference, since the intention is that death is a good thing.[61] If, however, one intends only to relieve suffering, and death is a byproduct of that decision through the use of higher and higher doses of pain medication (double-effect euthanasia), then there is a morally relevant distinction between killing and letting die. Moreover, the former intention is less "approved" than the latter, since in the latter instance the death occurs through natural causes and not by physicians assuming dominion over the lives of others.

Most thinkers agree that physicians today are not concerned with the mercy killing of vulnerable populations, as were the Nazis. Instead they are concerned about individuals who have made a judgment that continued life is no longer of use to them. In such instances, as James Rachels has argued in his classic article on the subject, there seems to be little moral difference between allowing the patient to die by the underlying disease process for humane reasons or directly ending the patient's life for humane reasons (presumably, too, providing the means for the patient to end his or her life).[62] Cessation of treatment, if intended to end a person's life, is permitted under most circumstances. Yet, how does this differ from direct killing for the same humane reasons? If there is little or no moral difference between these two actions, then surely physician-assisted suicide would make moral sense. As Judge Breck noted in his ruling about Dr. Kevorkian's assisted suicides, there is no legal distinction between a doctor withdrawing life support and a doctor injecting a patient with lethal drugs. "In both cases, the physician is causing death to occur."[63]

Arguments For and Against Physician-Assisted Suicide

A summary of the best arguments for and against physician-assisted suicide may now be offered.

Arguments in Favor of Physician-Assisted Suicide

The goal of medicine is to address the suffering of patients.[64] As that suffering increases, we have a duty to provide adequate pain control and to be concerned for the values of the patient in other areas that affect suffering. When pain or suffering become intractable to relief, however, and the patient repeatedly requests termination of life, then the physician has a duty from both the goal of medicine and from a wider responsibility of care, to provide relief through a lethal dose of medication.

Responsbility for technology also exists. In the old days, persons died at home without much medical intervention since little could be done. They often died prior to the end-stages of terminal illness; usually they died of an infection. Today, however, as a direct result of efforts to obtain greater quality of life during the terminal illness, people live longer but suffer more. Physicians who helped the patient gain some "additional time" through interventions such as chemotherapy must now take responsibility for those interventions and help patients to die in peace.

Independent of these considerations, physicians ought to respect individual autonomy. If rational suicide is possible, not only among the dying, but also by persons faced with advanced age, or even in the prime of life, then request for physician-assisted suicide should be honored,[65] given the proper precautions noted below, principally and even solely on the basis of an individual's assessment of the quality of his or her own life.

Concern for misuse of the powers of life and death is appropriate. Given the history of Nazi Germany and the broadening of powers during times of crisis, proper precautions must be established. But these procedures do not represent an impossible hurdle. The criteria in the Netherlands by which physicians are not prosecuted for assisting the dying of their patients (even though this assistance can be illegal)[66] can become legal procedures in the United States and elsewhere. Since physician-assisted suicide is an extraordinary and irreversible treatment, Drs. Quill, Cassel, and Meier noted that "it is not idiosyncratic, selfish, or indicative of a psychiatric disorder for people with an incurable illness to want some control over how they die."[67] They proposed the following clinical criteria that closely parallel the Dutch ones:

- The patient has an incurable condition associated with severe and unrelenting suffering. The patient understands this condition, the prognosis, and the types of comfort care available.
- The patient's suffering is certified by the primary physician as not being caused by inadequate comfort care.
- The patient must clearly and repeatedly request to die, of his or her own initiative.
- The physician must insure that the patient's judgment is not distorted.
- The assisted suicide must be carried out only in the context of a meaningful doctor–patient relationship.

- Consultation with a second physician is required to ensure that the patient's request is voluntary and rational, that the diagnosis is adequate, and that all other avenues of comfort care have been exhausted.
- Clear documentation of each condition is required.

In England, in response to the Cox case I noted above, Helme and Padfield proposed that a patient or nearest relative may apply for euthanasia or assistance. The doctor would register his or her intent to act, and disputes could be resolved through "euthanasia tribunals."[68]

Physician-assisted suicide has a long history in the privacy of the doctor–patient relationship. There is no need to consider it a "new" or even "experimental" idea. Quill asked, "I wonder how many families and physicians secretly help patients over the edge into death in the face of such severe suffering. I wonder how many severely ill or dying patients secretly take their own lives, dying alone in despair?"[69] In the past, physician aid-in-dying was done as a compassionate act very infrequently, when all other efforts to assist the patient failed. Today, because of public exposure and the legal risks that surround such an act, legalization is necessary. Furthermore, giving assistance, as Dr. Quill did to Diane, meant that he could not be present during her dying, nor could family members who might be legally implicated.[70] Legalization of the act would provide the goal of good terminal care, since caregivers and loved ones could be present.

The argument against physician-assisted suicide that tends to ground all respect for life in religious values about the sanctity of life, stewardship of life, and limits on human autonomy tends to assume that religious values would rule out taking the lives of patients. This is not entirely true. There is a diversity of religious arguments about euthanasia and assisted suicide reflecting a pluralism of religious viewpoints, even within the same traditions, in our culture.[71]

Arguments Against Physician-Assisted Suicide

First, the American Medical Association has ruled out any mercy killing, which is defined as "the intentional termination of the life of one human being by another."[72] Since a physician providing the means for helping another person commit suicide would be intentionally participating in bringing about the death of another, physicians should not participate in such actions. A stronger form of this argument is that no matter what society may encourage or permit, physicians ought not to kill on principle alone, as Willard Gaylin, Leon Kass, Mark Siegler, and Edmund D. Pellegrino have argued, since their role-specific duty is to heal rather than to end lives. As they stated, " . . . at least since the Oath of Hippocrates, Western medicine has regarded the killing of patients, even on request, as a profound violation of the deepest meaning of the medical vocation . . . Neither legal tolerance nor the best bedside manner can ever make medical killing medically ethical."[73]

Second, once the habits of taking dominion over the lives of others are formed, a slippery slope will develop in hard times, such that physicians will become more insistent in their role in assisting suicide, and begin to offer and urge it on patients who have become not only depressed about their circumstances, but also a burden to themselves and to others, even an economic burden. Surely some of this concern revolves around Kevorkian's public pronouncements and actions.

Another reason for hesitancy about legalizing physician assisted suicide is the tendency of physicians to become messianic about their role, such that gradual abuse might creep in under the guise of social, public, physician, and peer pressure to end one's life. Still another reason for caution might be that judicial case-by-case review of now-legal acts might not provide the same sort of weight as does the Dutch method of judicial review of illegal acts. No one in society can say that the action is legal, or suggest that it should be done, or encourage persons to consider it. The opposite would be true if the action were legalized.

Third, the doctor–patient relationship itself depends on trust, and if the public begins to mistrust the profession of medicine, because of its unhealthy participation in death-dealing, then the profession of medicine itself will suffer irreparable harm. Trust will no longer be available for the healing process to occur.

Fourth, it is ironic that at the very time the movements towards patients' rights and patient autonomy have sought to curtail physician infringement and control over individuals, society might permit such physicians to have the ultimate power of life and death in their hands. The quest for legalization and professional acceptance of euthanasia and assisted suicide under the mantle of patient autonomy has actually led to a significant expansion of the autonomy of the physician.[74] This is a dangerous move, since the profession itself can assume monumental proportions of belligerence rather rapidly. It lacks sufficient attention to the subjective, humane, and caring elements, and too easily slips into objectifying human life and manipulating it.[75]

Fifth, given the moral, political, social, and caregiver ramifications, have all the possible alternatives to supporting the dying been exhausted?[76] Efforts can still be made to use better means for controlling pain and addressing suffering. Much common ground still remains for individuals on both sides of the issue to continue the quest for more humane care of the dying, without violating the ancient proscriptions against killing.

Conclusion

The Dutch use the term "medical decisions at the end of life" to cover both physician assisted suicide and euthanasia. This unusual nomenclature underlines their conviction that such actions are truly medical decisions, done in conjunction with patients, like all other medical decisions.[77] It also stresses the importance of considering physician-assisted suicide, and euthanasia, only within the context of a thorough and committed program of terminal care. Providing death as a means to escape from intractable suffering is seen as the last kind act of a personal engagement with the dying person. It is done usually by the general practitioner who knows the patient very well, or by another physician in the hospital who has been involved throughout a long disease process with pain control. The physician must know the patient's values very well. And of course, the act itself is still illegal and by design may remain so, so that the possibility of judicial review occurs at all times.[78,79]

Three initiatives for legalizing physician assisted suicide and euthanasia have failed in the United States by narrow margins. This has happened, even though an overwhelming majority, perhaps even more than 75%, of the populace favors some form of aid in dying.[80] Perhaps one of the reasons is that legalizing physician-assisted suicide and active, direct euthanasia place too much power in the hands of physicians. Surely one of the significant political and ethical movements of our time has been the rise of patient's rights. In part this movement stems from a desire to check the enormous power of the modern medical establishment over the lives and values of human beings. Those who underline the importance of autonomy in bioethics in promoting the right of patients to ask for assistance in death, might fail to see that at least part of the autonomy movement has not arisen so much from a desire for absolute self-determination as from the desire to control physician dominance in health care decisions.

The most important point to consider, however, is that our society should spend some major effort on helping persons more completely during their dying by providing first-class terminal care based on the principles of hospice. This will require a thorough educational program in care of the dying, something which has not been offered in our professional schools to date.[81] Because denial of death is such a powerful force in a technologically equipped hospital environment, a terminal care program there has virtually no chance to succeed without an educational program.[82]

Further, without such a terminal care program, it appears that Americans might opt for physician-assisted suicide, as well as euthanasia, as a means of bypassing such a personal commitment to care for the dying. Sharon Selib Epstein, a social worker, voiced this concern very well. Reflecting both on her mother-in-law's death, and on the fact that in the Netherlands, persons in their 70s and 80s are far less likely to request aid in dying than their younger counterparts, Epstein said:

> All these experiences lead me to ask wider questions about death. Do we have a right to practice euthanasia with the excuse that we are ending the dying person's suffering? Might we perhaps be using euthanasia to end our own suffering? . . . Much more attention should be directed to the dying person in spite of the pain it gives us.[83]

Given the moral, professional, and social problems with physician-assisted suicide it might be better at this time to encourage physicians to be more engaged with the dying patient as a person, to develop training programs in medical schools for caring for the dying, and to thoroughly train all those who care for dying patients in the art of pain control. Further, society should guarantee terminal care for all citizens so that they need not fear bankrupting their family during their dying process.

Physician-assisted suicide might still be offered by caring physicians to patients for whom none of the above measures are adequate to relieve their suffering. This has been done "at the boundaries of life and death" for centuries. Howard Brody considered a similar idea, noting that the dispute between supporters and objectors to physician-assisted suicide will not be resolved on philosophical grounds because the arguments seem evenly balanced on both sides. Both sides must live with the tensions this creates, and physician-assisted suicide should be seen as a compassionate response to one sort of medical failure rather than as something to be prohibited outright or something to be established as general policy.[84] We should not rush to legalize such acts in place of the hard but rewarding work of providing proper terminal care.

References

[1]Kevorkian cleared of murder charge, (1992) *Chicago Tribune,* July 22, Sec. 1, 3.
[2]Anonymous (1992) Kevorkian helps another woman commit suicide, *Chicago Tribune,* Nov. 29, Sec. 1, 5.
[3]*Ibid.*
[4]*Ibid.*
[5]*See Hospital Ethics* **8,** no. 2 (March/April, 1992).
[6]*Ibid.*
[7]Anonymous (1992) 2 more assisted suicides before Governor OKs ban, *Chicago Tribune,* Dec. 16, Sec. 1, 6.
[8]*Ibid.*
[9]Margaret P. Battin (1982) *Ethical Issues in Suicide,* Prentice-Hall, Englewood Cliffs, NJ.
[10]Glenn C. Graber (1981) The rationality of suicide, in S. Wallace and A. Eser (eds.), *Suicide and Euthanasia: The Rights of Personhood,* University of Tennessee Press, Knoxville, TN.
[11]Richard Brandt (1975) The morality and rationality of suicide, in James Rachels (ed.), *Moral Problems,* Harper and Row, New York.
[12]Jacques Choron (1972) *Suicide,* Scribner's Sons, New York, p. 100.
[13]C. G. Prado (1990) *The Last Choice: Preemptive Suicide in Advanced Age,* Greenwood Press, Westport, CT.
[14]Erik Erikson (1963) *Childhood and Society,* 2nd ed., W. W. Norton, New York, p. 268.
[15]Harry R. Moody (1992) *Ethics in an Aging Society,* The Johns Hopkins University Press, Baltimore and London, p. 86.
[16]Tribune article. Also NEJM article.
[17]Connie Zuckerman (1992) A matter of consideration, cooperation and the courts: The tragic case of Jean Elbaum, *Precepts: Division of Humanities in Medicine Newsletter* **4** (3) (Nov.), 3, 6,7.
[18]*Ibid.,* 7.

[19]Albelt E. Gunn (1991) Risk–benefit ratio: The soft underbelly of patient autonomy, *Issues Law Med.* **7**, no. 2 (Fall), 139–154.

[20]David C. Thomasma (1992) Models of the doctor–Patient relationship and the ethics committee: Part one, *Cambridge Q. Healthcare Ethics* **1**, no. 1 (Winter), 11–32.

[21]Edmund D. Pellegrino and David C. Thomasma (1981) *A Philosophical Basis of Medical Practice,* Oxford University Press, New York.

[22]David C. Thomasma (1983) LEAD ARTICLE: Limitations of the autonomy model for the doctor–patient relationship, *Pharos* **46** (Spring), 2–5.

[23]David C. Thomasma (1985) Philosophy of medicine in Europe: Challenges for the future, *Theor. Med.* **6** (Feb.), 115–123.

[24]David C. Thomasma (1985) Editorial: Philosophy of medicine in the USA, *Theor. Med.* **6**, 239–242.

[25]David C. Thomasma and E. D. Pellegrino (1987) Challenges for a philosophy of medicine of the future: A response to fellow philosophers in the Netherlands, *Theor. Med.* **8**,187–204.

[26]Edmund D. Pellegrino and David C. Thomasma (1987) The conflict between autonomy and beneficence in medical ethics: Proposal for a resolution, *J. Contemp. Health Law Policy* **3**, 23–46.

[27]H. Tristram Engelhardt, Jr. (1987) *The Foundations of Bioethics,* Oxford University Press, New York.

[28]Nigel M. de S. Cameron (1992) *The New Medicine: Life and Death After Hippocrates,* Crossway Books, Wheaton, IL.

[29]Eric Cassell (1993) The relief of suffering, *Arch. Intern. Med.* **143** (March), 522–523.

[30]Anonymous (1992) Missouri and Kevorkian continue to provoke controversy, *Hospital Ethics* **8**, no. 6 (Nov./Dec.), 9.

[31]Timothy Quill (1991) Death and dignity: A case of individualized decision making,*New Engl. J. Med.* **324**,no. 10 (March 7), 691–694.

[32]D. Brahams (1992) Euthanasia: Doctor convicted of attempted murder, *Lancet* **340**, no. 8822 (Sept. 26), 782,783.

[33]Stanley M. Rosenblatt (1992) *Murder or Mercy: Euthanasia on Trial,* Prometheus Books, Buffalo, NY.

[34]Leon R. Kass (1992) "I will give no deadly drug." Why doctors must not kill, *Am. Coll. Surg. Bull.* **77**, no. 3 (Mar.), 7–17.

[35]Pellegrino and Thomasma, *For the Patients Good.*

³⁶Erich Loewy (1989) The restoration of beneficence, *The Hastings Center Report* **19**, 42,43.
³⁷W. Gaylin, L. Kass, E. D. Pellegrino, and M. Siegler (1988) Commentaries: Doctors must not kill, *JAMA* **259** (14) (8 April), 2139,2140.
³⁸L. Kass (1989) Arguments against active euthanasia by doctors found at medicine's core, *Kennedy Inst. Ethics Newsletter* **3** (Jan.), 1–3 and 6.
³⁹J. Muller and B. Koenig (1988) On the boundary of life and death: The definition of dying by medical residents, M. Lock and D. Gordon (eds.), *Biomedicine Examined,* Kluwer Academic, Dordrecht/Boston, pp. 351–374.
⁴⁰S. Braithwaite, D. C. Thomasma (1986) New guidelines on foregoing life-sustaining treatment in incompetent patients: An anti-cruelty policy, *Ann. Intern. Med.* **104**, 711–715.
⁴¹Ivan Illich (1976) *Medical Nemesis: The Expropriation of Health,* Pantheon Books, New York, p. 106.
⁴²*Ibid.*, p. 154.
⁴³David C. Thomasma (1990) The ethics of caring for vulnerable individuals, in *Reflections on Ethics,* American Speech-Language-Hearing Association, Washington, DC, pp. 39-45.
⁴⁴R. N. Proctor (1992) Nazi doctors, racial medicine, and human experimentation, in George Annas and Michael A. Grodin (eds.), *The Nazi Doctors and the Nuremberg Code,* Oxford University Press, New York, p. 23.
⁴⁵Jack Kevorkian (1991) *Medicide.*
⁴⁶Proctor, *Nazi Doctors,* p. 23-27.
⁴⁷R. N. Proctor (1988) *Racial Hygiene: Medicine Under the Nazis,* Harvard University Press, Cambridge. MA, pp. 179–189.
⁴⁸Stephen Post (1990) Severely demented elderly people: A case against senicide, *J. Am. Geriat. Soc.* **38**, no. 6 (June), 715–718.
⁴⁹David C. Thomasma (1992) Mercy killing of elderly people with dementia: A counterproposal, in Robert Binstock, Stephen Post, and Peter Whitehouse (eds.), *Dementia and Aging: Ethics, Values, and Policy Choices,* The Johns Hopkins University Press, Baltimore, MD, pp. 101–117.
⁵⁰Elie Wiesel (1992) Preface, in George J. Annas and Michael A. Grodin (eds.), *The Nazi Doctors and the Nuremberg Code,* Oxford University Press, New York), vii.

[51]*Ibid.*

[52]George Annas and Michael Grodin, Introduction, *The Nazi Doctors and the Nuremberg Code,* p. 3.

[53]Robert Proctor, Nazi doctors, racial medicine, and human experimentation, *in The Nazi Doctors and the Nuremberg Code,* pp. 17–31.

[54]*Ibid.,* p. 24.

[55]*Ibid.*

[56]As quoted at the head of Proctor, Nazi doctors, racial medicine, and human experimentation, in *The Nazi Doctors and the Nuremberg Code,* p. 17.

[57]As quoted in David Scott, Euthanasia: America's next challenge to life, (1992) *The Evangelist,* Diocese of Albany, NY, 67, no. 4 (Dec. 3), 4.

[58]*Ibid., loc. cit.*

[59]*Issues Law Med..*

[60]*Ibid.*

[61]David C. Thomasma, and Glenn C. Graber (1991) *Euthanasia: Toward an Ethical Social Policy,* Continuum, New York.

[62]James Rachels (1975) Active and passive euthanasia, *New Engl. J. Med.* **292,** no. 2 (Jan. 9), 78–80.

[63]*Ibid.*

[64]Eric J. Cassell (1982) The nature of suffering and the goals of medicine, *New Engl. J. Med.* **306,** no. 11 (March 18), 639–645.

[65]Marvin Kohl (1992) Altruistic humanism and voluntary beneficent euthanasia, *Issues Law Med.* **8,** no. 3 (Winter), 331,342.

[66]Pieter Admiraal (1989) Justifiable active euthanasia in the Netherlands, in Robert M. Baird and Stuart E. Rosenbaum (eds.), *Euthanasia: The Moral Issues,* Prometheus Books, Buffalo, NY, pp. 125–128.

[67]T. Quill, C. Cassel, and D. Meier (1992) Care of the hopelessly ill: Proposed clinical criteria for physician-assisted suicide, *New Engl. J. Med.* **327,** no. 19 (Nov. 5), 1380–1383.

[68]T. Helme, and N. Padfield (1992) Safeguarding euthanasia, *New Law J.* Oct. 2, 1335,1336.

[69]Quill, *Death and Dignity,* 694.

[70]*Ibid, loc. cit.*

[71]C. S. Campbell (1992) Religious ethics and active euthanasia in a pluralistic society, *Kennedy Inst. Ethics J.* **2,** no. 3 (Sept.), 253–277.

[72]The House of Delegates of the American Medical Association, Dec. 4, 1973, as quoted in Rachels, 78.

[73]W. Gaylin, L. Kass, E. D. Pellegrino, and M. Siegler (1988) Doctors must not kill, *JAMA* **259**, no. 14 (April 8), 2139–2140.

[74]Jos V. M. Welie (1992) The medical exception: Physicians, euthanasia and the Dutch criminal law, *J. Med. Philosophy* **17**, no. 4 (Aug.), 419–437.

[75]*See* the objections raised to physician-assisted suicide and euthanasia in a Newsletter of Evangelical Churches, Legalized physician-assisted suicide, (1992) *Discernment* **1**, no. 3 (Fall).

[76]C. S. Campbell (1992) "Aid-in-dying" and the taking of human life, *J. Med. Ethics* **18**, no. 3 (Sept.), 128–134.

[77]David C. Thomasma (1993) *Cambridge Q..* **2**, no. 1 (Winter), forthcoming.

[78]J. K. M. Gevers (1992) Legislation on euthanasia: Recent developments in the Netherlands, *J. Med. Ethics* **18**, no. 3 (Sept.), 138–141.

[79]Gerrit K. Kimsma (1992) Clinical ethics in assisting euthanasia: Avoiding malpractice in drug application, *J. Med. Philosophy* **17**, no. 4 (Aug.), 439-443.

[80]Helga Kuhse and Peter Singer (1992) Euthanasia: A survey of nurses' attitudes and practices, *Austral. Nurses J.* **21**, no. 8, 21,22, note that the same percentage occurs among Australian citizens, about 76% of those surveyed supporting allowing physicians to offer lethal doses of medication on request to "hopelessly ill" patients "in great pain." Sixty-six percent of nurses surveyed in the article said that patient had asked them to directly kill them at some time in their practice at least once, and 85% said that they participated in ending a patient's life directly when asked to do so by a doctor.

[81]Anonymous (1991) Hospitals, physicians paying more attention to pain control, *Med. Ethics Advisor* **7**, no. 11 (Nov.), 133–137.

[82]William Winslade (1992) Teaching about dying, *Choice in Dying News* **1**, no. 4 (Winter),

[83]Sharon Selib Epstein (1992) What the dying give to the living, *Chicago Tribune,* December 11, Sec. 1, 21.

[84]Howard Brody (1992) Assisted death—A compassionate response to a medical failure, *New Engl. J. Med.* **327**, no. 19 (Nov. 5), 1384–1388.

Life, Death, and the Pursuit of Destiny

The Case of Thomas Donaldson

Evelyne Shuster

There is, in all of us, a certain part that lives outside of time.[1] We become aware of our mortality only at exceptional moments, a mild indisposition, a sudden handicap, or an illness that forces us to change our lifestyle and restrict our activities. But most of the time we feel immortal. First, there was a belief that if we obey nature's commands and follow its dictates, this would "free men from an infinity of maladies both of the body and of the mind, and possibly of all the infirmities of age,"[2] then a belief that technology itself could replace nature and transcend its very limits.[3] Today, death is safely removed to a comfortable distance. Sophisticated life-saving techniques constitute society's standing order against diseases. However, to fight death itself, has not usually been within the domain of medicine. Yet, this is the request Thomas Donaldson made to a California Court of Appeal when he sought a declaratory judgment that could give him the legal right to "mercy-freezing."[4]

The Case of Thomas Donaldson

Thomas Donaldson, a mathematician and computer software specialist from Santa Barbara, California, suffered from grade 2 astrocytoma, an inoperable brain tumor that progressively leads to a permanent vegetative state and death. Facing intolerable suffering and incapacitation, Donaldson wanted to die when he was still in possession of his full mental and physical capacities, and have his body cryogenically preserved in the hope of being brought back to life when a cure is found for his condition (premortem cryogenic suspension).

Cryonics or cryopreservation of all or part of human bodies (postmortem cryopreservation), although unusual, is currently available at a price.[5] On the patient's death, the body is prepared and frozen by special methods that prevent or minimize freezing injuries (cellular destruction by crystal formation). To date, cryopreservation of cells, gametes, tissues, and embryos has been successful. The freezing of complex vital organs, however, has failed to produce methods for allowing subsequent survival of thawed organs on transplantation. Total cryonic suspension and revival of the entire human body remains speculative at best.[5]

Premortem cryogenic suspension is different. Donaldson explained "when conventional treatment fails, he plans to go to a cryonics center and to end his life. Cryogenic scientists and technicians would anesthetize him, his temperature would be lowered, and he would be pumped full of solutions to prevent severe cellular destruction. His head would then be cut off, placed in a metal canister, and frozen. There, he would rest in peace in a bath of liquid nitrogen alongside other canisters."[6] To carry out his plans he needed "special advice and encouragement" of Carlos Montragon who, then, would supervise "over the cryogenic suspension process." Because assisted suicide is illegal, Donaldson and Montragon sought an injunction to protect Montragon from criminal prosecution, and to prevent the county coroner from examining Donaldson's remains.

Donaldson claimed a constitutionally protected right to privacy and a right to self-determination[4] (p. 3) paramount to any state interest in maintaining life. The trial judge ruled that Donaldson and Montragon failed to state a cause of action, and dismissed the case. They appealed.

The Appeal Court's Opinion

On appeal, Donaldson argued that judicial decisions have explicitly recognized the right of competent adults to refuse any or all unwanted medical treatment, under the informed consent doctrine. He contended that there is no medical, or ethical distinction between letting him die by honoring his right to refuse treatment (passive euthanasia), and actively causing his death at his request (active voluntary euthanasia). He quoted Joseph Flechter:

> *"Not doing anything is doing something.* It is a decision to act every bit as much as deciding for any other deed. If I decide not to eat or drink anymore, knowing what the consequence will be, I have committed suicide as surely as if I had used a gas oven."[4] (p. 5 emphasis added)

At the very least, Donaldson claimed he should have a right to receive, and Montragon a right to provide, "advice and encouragement" for his suicide.

To bolster his argument, Donaldson relied on Bouvia[7] where a California Appeals Court Judge stated, in his concurring opinion:

> "This state and the medical profession, instead of frustrating [Bouvia's] desire, should be attempting to relieve her suffering by permitting, and, in fact, assisting her to die with ease and dignity. [The right to die] should include *the ability to enlist assistance from others, including the medical profession, in making death as painless and quick as possible"* (p. 6, emphasis added)

Carlos Montragon claimed that he too had a constitutionally protected right, namely to free expression, and thus, should be permitted to counsel, advise, and encourage Donaldson in his suicide. In his words: "Penal Code 401 impairs the exercise of free speech. Since suicide is not illegal, the state may not prohibit free speech that encourages a lawful act"[4] (p. 8).

The court ruled that Donaldson's rights to privacy and self-determination are not absolute, and when life is at stake, these rights must be balanced against the four countervailing state interests in

1. The preservation of life;
2. The protection of innocent third parties;
3. The prevention of suicide; and
4. The protection of the ethical integrity of the medical profession.

The court reasoned that the state may decline to consider the quality of a particular human life and can assert an unqualified interest in life, as in Cruzan.[8] In so doing, the court maintained that the state does not violate a constitutionally protected liberty interest or privacy right. The court further ruled that significant differences exist between the withholding and the withdrawing of life-sustaining therapy, and a "self-infliction of deadly harm" or suicide:

> "While, in the case of treatment refusal, patients are seen as 'passive victims' of their underlying illnesses, in the case of 'self-infliction of deadly harm,' a person is actively being killed. If there is a similarity between the two cases, it is only in appearance. In life support termination, there is a fiction of medical determinism. Patients are seen as passive victims of their illness. They do not choose to die: death overtakes them. Their physicians do nothing to help them die. Death overwhelms them too"[4] (p. 5).

Thus, in the opinion of the court,

> "Donaldson is not asking for a permission to refuse life-
> sustaining treatment. What he is asking this court is to sanc-
> tion something quite different: the right of a third person to
> kill him with impunity. No statute or judicial opinion coun-
> tenances Donaldson's decision to consent to be murdered or
> to commit suicide with the assistance of others. This essen-
> tial dissimilarity between the right to decline medical treat-
> ment and any right to end one's life causes physicians to
> incur no criminal liability in the first case, and to be legally
> liable in the second one"[4] (p. 6).

The court recognized that Donaldson may have a "fundamental" right to take his own life and end his suffering. However,
it does not follow that he could involve others in the process with
impunity. The state interests in protecting the lives of those who
wish to live no matter the circumstances, and in ensuring that
people are not unduly influenced or encouraged to kill themselves take precedence over the individual. In the court's words:

> "The difficulties, if not the impossibilities, of evaluating the
> real motives of the assister, or determining the presence of
> undue influence, pressure or coercion make it necessary to
> outlaw all assisted suicides. [Thus], To conceive a particular
> judicial procedure that can supervise over a single case with-
> out endangering society as a whole is impossible"[4] (p. 6).

The court ruled that Donaldson has no constitutionally protected right to a "state assisted death." The legal and philosophical problems created by Donaldson's predicaments are legislative
matters; they are not judicial. The court thus decided: "To prevent
the coroner from doing his job requires legislative action rather
than judicial one"[4] (p. 6).

The court further decided that Penal Code 401 does not impair
Montragon's exercise of free speech: "[While] the constitutional
guarantees of free speech protect the individual's freedom to

speak, write, print or otherwise communicate information or opinion, it does not follow that Montragon's actions are also protected when he would have specifically intended Donaldson's suicide and have had a direct participation in bringing it about. Regulation of conduct bearing no necessary relation to the freedom to disseminate information or opinion is not constitutionally protected," and thus, Montragon enjoys no constitutional protection for his planned participation in Donaldson's suicide[4] (p. 7).

Discussion

The court rested its decision on the premise that there are essential differences between exercising one's right to refuse life-saving treatment and die, and requesting, and receiving assistance with death. In the United States, no statute or judicial opinion has ever endorsed physician-assisted suicide, or euthanasia (the medical administration of a lethal agent to end a patient's life), even when there is explicit and unequivocal consent.

Recently, the debate over euthanasia has intensified, when Jack Kevorkian, a Michigan retired pathologist, used his own devices and materials to assist in the death of sixteen individuals with terminal or painful illnesses.[9] Despite the shocking nature of these cases, Americans have sympathized with the plight of those who suffer the effects of a long and agonizing illness and want to end their lives. In 1991, Dereck Humphrey, the founder of the Hemlock Society, a group that has long advocated voluntary euthanasia, published a best-selling kill-thyself guide, *Final Exit*,[10] which received nationwide attention. In Washington State, *Initiative 119,* which would have legalized euthanasia (although unsuccessful) was supported by 44% of the voters. In California, the *Californians Against Human Suffering,* in 1988, and again in 1992, proposed a measure (Proposition 161) to legalize physician-assisted suicide and active euthanasia (the measure was narrowly defeated).[12] Similar proposals are under consideration by the legislature in Oregon, New Hampshire, Florida, and Michigan.[12]

Health professionals have also taken a position on physician-assisted death.[13] A survey of physicians indicates that 60% favor legalizing euthanasia, though nearly half of those in favor would refuse to perform it.[14] In 1991, a physician published a description of his involvement with the suicide of one of his patients, Diane.[15] In 1992, he and two other physicians called for a new public policy that would permit physician-assisted suicide (but not active euthanasia).[16] To an increasing number of individuals it has seemed unfair that patients—competent and incompetent—may refuse life-sustaining treatment and die, and yet they are "abandoned" by their physicians when there is nothing more medicine can offer.[16] In this context some have argued that "the medical profession's repeated and firm rejection of any participation by physicians in assisted suicide begins to appear self-serving in its emphasis on a professional scrupulosity that seems blind to the expressed needs of patients."[17]

Has the time come to reassess the physician's traditional role of fighting disease, preserving life, and alleviating suffering, and to include within that role purposefully ending the lives of patients at their requests? Can the principle of autonomy, which justifies a right to forgo any treatment, be expanded to include a right to euthanasia, and if so, what will the potential consequences be?

Informed Consent

Of all ethical principles—such as autonomy, justice, beneficence, and nonmaleficence—Americans seem mainly to value autonomy, which has been shaped by a unique concern for individuals. As a society, we have fashioned a concept that has come to support all kinds of rights—self-determination and personal privacy, freedom, and independence, liberty of choice and action, and control over decisions and activities. In health care (and in life) individuals enjoy the freedom to determine their preferences and to live them, so long as they do not harm others. Individuals, more often than not, seem to favor utilitarian values in the conduct of their lives, and to have adopted John Stuart Mill's often quoted proposition that "the only purpose for which power can be rightfully exercised over any member of a civilized community, *against his will,* is to *prevent harm to others. His own good, either physical or moral, is not a sufficient warrant*"[18] (emphasis added). Informed consent is predicated on the principle of autonomy and the concept of a person as a free moral agent whose values and preferences must be honored. Absent countervailing obligations, a physician has a duty to respect the right of autonomous patients to refuse any or all recommended medical treatment.

Informed consent, however, does not justify providing treatments that present a significant harm to patients. For instance, in criminal law, consent has never been recognized as a defense to such a crime as providing harmful, dangerous, or unacceptable treatment to a consenting adult.[19] Nor can patients demand treatments that, according to reasonable medical judgment, are medically unsound and detrimental to their well-being. If they do, the physician has a right, and indeed a duty, to refuse to provide such treatments despite the patient's specific request. The physician's affirmative duty is to administer sound medical treatments, i.e., those treatments which are likely to *do good,* or, at least, *do no harm.*[13] Soundness of medical treatment arises from the ethical principles of beneficence (doing good), and non-maleficence (doing no harm). Arguably, interventions which intentionally cause the death of a patient cannot constitute sound medical treatment, and thus, physicians have a duty *not* to provide them. Moreover, once a physician is involved in causing a patient's death, a decision to die is no longer strictly individual. The involvement of a second person must implicate that person's autonomy as well as societal norms. In this context the welfare of society could take precedence over the individual to bring about the greatest happiness for the greatest number."[13] Traditionally, the physician's role has been to heal. To give physicians the power and authority to kill patients under the informed consent doctrine could not only undermine this role, but also create a potential for abuse (as the Nazi atrocities tragically illustrate), and thus, cause more harm than good, everyone considered.

A right creates a corresponding duty to respect such a right. Should a moral (or legal) right to euthanasia be recognized, it would create a corresponding moral (or legal) duty to kill. No ethical (or legal) system has thus far justified such a duty, except in the unusual circumstances of capital punishment, self-defense, and war. To this date, society has been reluctant to include another exception to its ultimate moral imperative: *You shall not kill.*

The Netherlands Experience

The above arguments have thus far been strong enough to prevent a societal sanction of euthanasia. Essentially, they hold that the rights of patients must be respected, but that they may not overrule societal welfare. If euthanasia were to be permitted, society would not be better off because it would favor killing patients rather than looking for better and new ways to confront death through better hospices, pain control management and comfort care. In a society where hopelessly ill patients are often perceived as expendable,[20] euthanasia would foster the view that dying patients are "abnormal," or "subhuman," and further encourage them to "volunteer" for questionable treatment, or to "consent" to euthanasia simply because there is nothing more that can be done for them and to them. Since physicians who are willing to assist in the death of their patients must, at least in principle, agree with them that their lives are not worth living, these physicians could apply this value judgment to patients in similar conditions, and believe that they, too, should consent to be killed. The poor, the very sick, the disenfranchised, and those who do not engage our sympathy could be led to think that their lives are not worth living, and thus, that they have a "duty to die."[22] Presumed consent to euthanasia could be the next logical step where hopelessly ill, incompetent patients have no surrogate and/or advance medical directives. As a result, physicians, not patients, would be empowered to make life and death decisions.[22]

The Netherlands experience is an example. The Netherlands is the only country in the world that officially tolerates euthanasia. The first national study of the practice indicates that euthanasia occurs in 1.8% of all deaths, and assisted suicide in 0.3%.[23] The study also indicates that in 0.8% of cases, lethal drugs were administered to incompetent patients with physicians acting as surrogates. Dutch physicians have not been concerned with this high rate of involuntary active euthanasia. The study illustrates an erosion of the well-established criteria for euthanasia that specify that the decision to die must be made explicitly and repeatedly by a well-informed, free, and competent person. By refusing to prosecute physicians who administer lethal agents to patients who were not competent to request them, the Dutch society has empowered them to kill with virtual impunity.[24]

Cryogenic Treatment: A Compelling Case of Physician-Assisted Death?

Donaldson's claim, nonetheless, creates different problems. He has compellingly argued that the purpose of his planned death is to live, *not to die,* and thus, to act *in conformity with the state interest in life.* Just as advance life support techniques can prolong life, so too, cryogenic suspension could ensure his survival, perhaps even bring about longevity that verges on "immortality."[6] On this basis, Donaldson requested the court to make an exception to the law. The court, however, decided to ignore the issue of cryonic "treatment," and to frame the request in terms of a classic example of active euthanasia. By avoiding, or choosing to ignore, the unusual aspect of this case, the court's long discussion of differences between passive and active euthanasia contributes little to further the debate, much less to provide a useful opinion on this controversial issue. Indeed, the use of cryonic treatment raises several questions: Could assisted suicide and cryogenic suspension be a means to the end of living, and a "treatment" of choice when there are reasons to believe that the patient's life can be saved? Where could the state get its power to overrule a competent patient's right to consent to life-saving treatment, such as cryonics? Can the state justify forcing hopelessly ill patients to live a life of prolonged suffering? To be sure, society has been prompt at condoning all kinds of "heroic" measures (animal organs, such as pig liver and baboon heart for transplant, artificial heart implant), when the rationale has been presumably to "save patients' lives."[21] Under this assumption, physicians subject dying patients to extreme, and (questionable) experimental treatments, when there is no scientific or medical evidence of benefit.[21]

Yet, it is notable that society's deep commitment to saving lives through scientific progress and medical technology has not included cryonic treatment. This could be because the implications appear so far beyond reason that, even if cryonic treatment were possible, it would be difficult to justify it, philosophically, spiritually, morally, and socially.[5] For example, if death could be permanently avoided, should it be done? Would it be fair to use scarce medical resources to extend life in such a way? Should medicare or medicaid pay for this "high technological death?" Who may decide who should be cryopreserved and who should not? How, and on what basis, should the reintroduction of humans into a different society be planned? Would society accept the extra cost of caring for impaired persons who poorly survive the freezing process, whereas others may be denied access to treatment because of cost? If access to cryonic treatment is limited, would such factors as social status, value, or utility influence the decision about who shall live and who shall die? How could accountability be maintained in case of negligence? What, if any, would the rights of a "frozen person" be, and how would these rights be negotiated? Who "owns" a frozen person?

Cryonic treatment could also change the physician–patient relationship. Patients would no longer appear to have a right to refuse treatment that could ensure survival, since it would be contrary to the state interest in life. Physicians also would seem to have a duty to treat using cryonics. If they refuse, they might be held liable for murder. Logically, if science will ultimately provide a cure for all diseases, including freezing damages, a case could be made for the cryopreservation of embryos (currently possible) until a cure is found for all diseases. This logic of survival at all cost would obviously have its limits, if it were not preposterous.

Donaldson, nonetheless, has other options. He may decide against the court to implement his planned death. Even if Montragon is actively involved in his death, there is no certainty that he would be prosecuted.[25] The existence of criminal liability in assisted suicide (or euthanasia) is no guarantee that those who kill patients (usually family members or friends) will be prosecuted, or that prosecution will be successful. Prosecution has been infrequent. Had there been prosecution, rarely did those involve in death decisions spent any time in jail.[25]

Physician-assisted suicide creates different problems when it is argued that the involvement of a physician stems from advances in medical technology that permit treatment (not cure) of conditions once considered fatal.[17] The rationale is that, once the medical profession brings patients to a state of extended suffering (chemotherapy, radiation therapy, xenografts, artificial heart devices), it may not be good medical practice for physicians to abandon them there.[17] This argument, however, is unconvincing because competent patients can always refuse life-sustaining treatment and die. It is often when patients consent to the latest medical technique to prolong life from an expected few months to years that the question of euthanasia may arise. The responsibility of these decisions rests on patients (not physicians) and their willingness to consent to life-prolonging therapy. Patients who have the mental and physical capability to end their lives have no reason to involve physicians except if they want to defer responsibility by making suicide a medical problem, or to transfer responsibility into a "murky realm that permits everyone to be blameless."[9] This may not come as a surprise in an era when almost all human problems are perceived as medical problems to be addressed within the domain of medicine.

The medicalization of suicide may be further enhanced by a need to control health care expenditures, and to spend health care dollars on patients who could get well before those who have incurable diseases.[26] For example, in Oregon, a survey of 576 randomly sampled health care executives indicates that more than half believe that "given an incurable illness, euthanasia should be considered as *a basic service if the patient wants it*[27] (emphasis added). However, there have been no serious public debates on this issue, which may explain the discrepancy between public opinion polls that seemingly favored legalized euthanasia, and what, in fact, is happening in California, Washington, and other states.

Conclusion

Medical culture reflects a broader culture. Along with a powerful emphasis on individual autonomy, and a broad range of rights, Americans value progress through technological means. "We, Americans, are made to be masters of destiny, not victims of fate."[28] American society is a technologically driven, action-oriented society. These values extend to medical technology through a deep commitment to a belief that scientific progress can provide even greater power over disease. Technological dependency is also tied to a tendency to "look always for the easiest solutions to complex problems."[29]

To be sure, a comprehensive health care system that guarantees access to needed services, more and better hospices, better pain control management, and better comfort care of the dying patient may reduce the need for euthanasia. However, public debates would have to demonstrate that this need is *real,* and determine, under what conditions, if any, a physician-assisted death is justified. Absent such a debate, there can be no justification for legalizing euthanasia based on current medical, ethical, and societal arguments, because these arguments have consistently been fairly distributed on both sides of the issue.

On the other hand, we, as a culture, do far better in the application of a "technological fix" than in building social arrangements (e.g., home care, hospices, comprehensive comfort care of the dying patient) that must be sustained over time in coping with expensive, frustrating, and often intractable problems."[30] "Romancing technology" is appealing in a culture that is one of action verb: to act, to intervene, to treat, to improve life, and to forestall death. Ultimately, it may be that the public will opt for action and the direct involvement of physicians in death decisions, *only because* action appears to be easier than inaction, i.e., the passive acceptance of fate, in addressing the one existential problem, death. And thus, as the Donaldson case illustrates, by recognizing their rights to euthanasia, individuals may mistakenly think they have become "masters of their destinies."

Notes and References

[1]Kyndera, M. (1992) *Immortality,* Harper Perennial, New York, pp. 1-4.
[2]Descartes, R. (1973) *Discourse on Method,* part VI, p. 122, in, *The Philosophical Works of Descartes,* vol. 1, Haldane and Ross trans., Cambridge University Press, Cambridge, UK.
[3]Shuster, E. (1972) *The Self-Medicated Man,* Presses Universitaires de France, Paris.
[4]*Thomas Donaldson et al.* v. *John Van De Kamp et al.* (2d. Civil No. B055657). 1992 Cal. App. Lexis 104; 4 Cal. Rptr. 2d 59.
[5]Goldman, A. J. (1978) Cryobiology, in *Judaism Confronts Contemporary Issues,* Shengold, New York, pp. 151–157; *see also,* Panuska, J. A., (1978) *Cryonics,* in *Encyclopedia of Bioethics,* vol. 1, Reich, W. T., ed., Free, New York, pp. 216–219.

[6]Goodage, Maria (1990) Manpins his hopes on a frozen future. *USA Today,* Sept. 25, 6A.

[7]*Bouvia* v. *Superior Court,* 1986, 179 Cal. App. 3d 1127.

[8]*Cruzan* v. *Director, Mo Health Dept.* 1990 497 US 111 L. ed. 2d 224, 110 S.Ct.

[9]Auburn, H. (1992) Doctor Assists Two More Suicides in Michigan, *New York Times,* Dec.16, A21; *see also,* Gelman, D., Springen, K. (1990) The Doctor's Suicide Van, *Newsweek,* June 18, pp. 46–49; Katherine S. (1992) Dr. Death's Clients: All Women, But Why? *The Philadelphia Inquirer,* Dec. 20; C-1, C-3; Gibbs, N. (1993) Rx for death, *Time,* May 31; 35–39.

[10]Humphry, D. (1991) *Final Exit.* The practicalities of self-deliverance and assisted suicide for the dying, The Hemlock Society, Eugene, OR.

[11]Misbin, R. I. (1991) Physicians' aid in dying, *New Engl. J. Med.* **325,** pp. 1307–1311.

[12]Gianelli, D. M. (1991) A right to die: debate intensifies over euthanasia and the doctors role, *Am. Med. News,* Jan. 7, pp. 9–18; Metz, B. (1991) Despite defeat of state's suicide initiative, issue still unsettled, *Am. Med. News,* Nov. 18, 29; Reinhold, R. (1992) California to decide if doctors can assist in suicide, *New York Times,* Friday, Oct. 9. A-1.

[13]Council on Ethical and Judicial Affairs, American Medical Association, (1992) Decisions near the end of life. *JAMA* **267,** 2229–2233; *see also,* Wanzer, S. H., Federman, D. D., and Adelstein, S. J., et al. (1989) The physician's responsibility toward hopelessly ill patients, *New Engl. J. Med.* **320,** 844–849.

[14]Angell, M. (1988) Euthanasia, *New Engl. J. Med.* **319,** 1348–1350.

[15]Quill, T. E. (1991) Death and dignity: A case of individualized decision making, *New Engl. J. Med.* **324,** 691–694.

[16]Quill, T. E., Cassell, C. K., Meier, D. E. (1992) Care of the hopelessly ill—Proposed clinical criteria for physician-assisted suicide, *New Engl. J. Med.* **327,** 1380–1384.

[17]Cassell, C. K., Meier, D. E. (1990) Morals and moralism in the debate over euthanasia and assisted suicide, *New Engl. J. Med.* **323,** 350–352; Benrubi, G. I. (1992) Euthanasia—The need for procedural safeguards, *New Engl. J. Med.* **326,** 197,198; Brody, H. (1992) Assisted death—A compassionated response to a medical failure, *New Engl. J. Med.* **327,** 1384–1388.

[18]Stuart Mill, J. (1978) *On Liberty,* Hackett, Indianapolis, IN, 74.

[19]When doctor has AIDS, (1991) *The National Law Review*, Sept. 9.

[20]Brauer R. (1988) The promise that failed, *The New York Times Magazine*. Sunday, Aug. 28, 35.

[21]Annas, G. J. (1988) Baby Fae: The "anything goes" school of human experimentation, in *Judging Medicine,* Humana Press, Clifton, NJ, 384–390; Kolata, G. (1989) Critics fault secret effort to test AIDS drug, *New York Times,* Tuesday, Sept. 19; Avis sur les Expérimentations sur les Malades en Etat Végétatif Chronique, in Lèttres d'Information du Comité Consultatif National d'Ethique pour les Sciences de la Vie et de la Santé, April 1986, A.

[22]Goodwin, J. S. (1991) Mercy killing: Mercy for whom? Piece of my mind. *JAMA* 265, 3, 325; How did Germany's doctors become torturers? (1991) *Boston University News,* Nov. 4–10.

[23]Van der Mass, P. J., Van Delden, J. J., Pijnenborg, L., et al. (1991) Euthanasia and other medical decisions concerning the end of life, *Lancet* **338,** 669–674.

[24]Gomez, F. C. (1991) *Doctors and Dying,* Regulating death: the case of the Netherlands, Free, New York.

[25]Glantz, L. H. (1987) Withholding and withdrawing treatment: The role of the criminal law. *Law, Medicine, and Health Care* **88,** 231–241.

[26]Callahan, D. (1990) *To kill and to ration:* Preserving the difference, in *What Kind of Life,* Simon and Schuster, New York; Dougherty, C. J. (1991) Setting health care priorities, *Hastings Center Report,* May–June 1–9.

[27]Kenneth V. (1990) health care executives and medical ethics, *Hastings Center Report,* Nov./Dec., 2,3.

[28]*Ronald Reagan: Keynote Address.* Republican National Convention, Houston, Texas, *New York Times,* Tuesday, Aug. 22, 1992, A3.

[29]Blank, R. H. (1988) *Life, Death, and Public Policy,* Northern Illinois University Press **11.**

[30]Mechanic, D. (1986) *From Advocacy to Allocation:* The evolving American health care system **207,** Free, New York.

Index